Iris Gebauer

An Anisotropic Model for Galactic Cosmic Ray Transport

Iris Gebauer

An Anisotropic Model for Galactic Cosmic Ray Transport

and its Implications for Indirect Dark Matter Searches

Südwestdeutscher Verlag für Hochschulschriften

Impressum/Imprint (nur für Deutschland/ only for Germany)
Bibliografische Information der Deutschen Nationalbibliothek: Die Deutsche Nationalbibliothek verzeichnet diese Publikation in der Deutschen Nationalbibliografie; detaillierte bibliografische Daten sind im Internet über http://dnb.d-nb.de abrufbar.

Verlag: Südwestdeutscher Verlag für Hochschulschriften Aktiengesellschaft & Co. KG
Dudweiler Landstr. 99, 66123 Saarbrücken, Deutschland
Telefon +49 681 37 20 271-1, Telefax +49 681 37 20 271-0
Email: info@svh-verlag.de
Zugl.: Karlsruhe, KIT, Diss., 2010

Herstellung in Deutschland:
Schaltungsdienst Lange o.H.G., Berlin
Books on Demand GmbH, Norderstedt
Reha GmbH, Saarbrücken
Amazon Distribution GmbH, Leipzig
ISBN: 978-3-8381-1529-0

Imprint (only for USA, GB)
Bibliographic information published by the Deutsche Nationalbibliothek: The Deutsche Nationalbibliothek lists this publication in the Deutsche Nationalbibliografie; detailed bibliographic data are available in the Internet at http://dnb.d-nb.de.

Publisher: Südwestdeutscher Verlag für Hochschulschriften Aktiengesellschaft & Co. KG
Dudweiler Landstr. 99, 66123 Saarbrücken, Germany
Phone +49 681 37 20 271-1, Fax +49 681 37 20 271-0
Email: info@svh-verlag.de

Printed in the U.S.A.
Printed in the U.K. by (see last page)
ISBN: 978-3-8381-1529-0

Contents

List of Abbreviations

ADI	Alternating Direction Implicit (Method)
AGN	Active Galactic Nucleus
aPM	Anisotropic Propagation Model
B/D	Bulge-over-Disk (Ratio)
CR	Cosmic Ray
CMB	Cosmic Microwave Background
CO	Carbonmonoxide
DM	Dark Matter
DMA	Dark Matter Annihilation
EAS	Extensive Air Shower Array
EB	Extragalactic Background
GC	Galactic Center
GZK	Greisen-Zatsepin-Ku'zmin (Cutoff)
H_2	Molecular Hydrogen
HI	Atomic Hydrogen
HII	Ionized Atomic Hydrogen, i.e. Protons
IC	Inverse Compton
LIS	Local Interstellar Spectrum
ISM	Interstellar Medium
ISRF	Interstellar Radiation Field
LKP	Lightest Kaluza-Klein Particle
LSP	Lightest Supersymmetric Particle
MC	Molecular Cloud
MCC	Molecular Cloud Complex

NFW	Navarro-Frenk-White
PISO	Pseudo-Isothermal (Profile)
SGRG	Soft-γ-Ray-Gradient (Problem)
SN	Supernova
SNR	Supernova Remnant
UV	Ultraviolet
WIMP	Weakly Interactig Massive Particle
X_{CO}	Ratio of Molecular Hydrogen to Carbonmonoxide Densities

List of Figures

List of Tables

Chapter 1

Introduction

The highest energies of single particles are currently not produced in particle accelerators on Earth, but in interstellar space. The Earth is under constant bombardment by cosmic rays, energetic particles, which come to us from outer space. These particles are measured either through satellites, balloons, or Earth based experiments. The origin of cosmic rays has been intriguing scientists since 1912 when Victor Hess carried out his famous balloon flight to measure the ionization rate in the upper atmosphere. Since then a number of important discoveries, like the muon, the pion and the kaon, amongst others, were made in cosmic rays. Today the study of cosmic ray physics brings together scientist from a variety of fields:

Astrophysicists are interested in the sources and acceleration mechanisms for cosmic rays. Together with astronomers they use Galactic cosmic rays as a probe to constrain the properties of the interstellar medium and the local environment of the Sun, as well as turbulent regions such as pulsar nebulae. The energy density of relativistic particles is about 1 eV cm^{-3} and is comparable to the energy density of the interstellar radiation field, magnetic field, and turbulent motions of the interstellar gas. This makes cosmic rays one of the essential factors determining the dynamics and processes in the interstellar medium. Particle physicists look for possible signals of dark matter annihilation in cosmic rays. With the increasing understanding of the history of our universe the necessity for dark matter and consequently a viable candidate becomes a pressing problem. Together with cosmologists particle physicists try to constrain the profile of the dark matter distribution in the Milky Way or dwarf galaxies in the Milky Way's halo through the observation of electromagnetic emission or charged stable decay products from dark matter annihilation. In the last decades indirect dark matter searches in Galactic cosmic rays have lead to a fruitful interplay of the above disciplines.

To extract the information which is contained in cosmic ray abundances and γ-ray fluxes one needs a model of particle production and propagation in the Galaxy. Though the basic features of particle diffusion in the Galaxy seem to be well-established, the continuous flow of new and more accurate data from space, balloon and ground based experiments motivates further development of models. Analytical and semi-analytical models are able to interpret one or only a few features and often fail when they try to deal with the whole variety of data. Therefore more realistic and consistent models are required which would be able to incorporate many processes and astrophysical data of many different kinds simultaneously.

The details of the specific physical mechanism that regulates the motion of cosmic rays are not yet known, but a decisive role is played by the galactic magnetic field. Because of the absence of a definite theory that explains the nature of the propagation of cosmic rays based on a rigorous picture of the interaction of charged relativistic particles with the interstellar medium, one uses approximate semi-empirical models. These models are derived from the basic properties of cosmic rays:

The spectrum of cosmic rays can be approximately described by a single power law with index -2.7 from ~ 10 GeV to the highest energies ever observed $\sim 10^{20}$ eV. The only feature observed below 10^{18} eV is a knee around 10^{15} eV, where the spectrum becomes slightly softer. Because of this featureless spectrum, it is believed that cosmic-ray production and propagation is governed by the same mechanism over decades of energy, the same mechanism at least works below the knee and the same or another one works above the knee. The sources of cosmic rays and in particular the acceleration mechanisms are not completely understood. They are believed to be supernovae and supernova remnants, pulsars, compact objects in close binary systems, and stellar winds, where charged particles are accelerated in shock waves. Observations of X-ray and γ-ray emission from these objects reveal the presence of energetic particles thus testifying to efficient acceleration processes near these objects. Particles accelerated near the sources to almost the speed of light propagate tens of millions years in the interstellar medium, as it is known from the relative abundances of radioactive instable isotopes the so-called "cosmic clocks". This means that cosmic rays have to be confined to the Galaxy by some mechanism, which is possible if cosmic rays continuously scatter on magnetic turbulences and as a consequence perform a random walk which can be modeled by diffusion. Possible scattering centers are the turbulent magnetic fields generated by the dilute plasma of cosmic rays themselves. During their time in the gaseous Galactic disk, cosmic rays can interact with the interstellar material where they lose or gain energy, their initial spectra and composition change, they produce secondary particles and γ-rays. The destruction of primary nuclei via fragmentation gives rise to secondary nuclei and isotopes which are rare in nature. The variety of isotopes in cosmic rays allows one to study different aspects of their acceleration and propagation in the interstellar medium as well as the source composition. Stable secondary nuclei tell us about the diffusion coefficient and reacceleration in the interstellar medium. Long-lived radioactive secondaries allow one to constrain global Galactic properties such as the cosmic ray flux towards the boundary.

All these together allow us in principle to build a model of particle propagation in the Galaxy. In the simplest case diffusion is assumed to be isotropic with the same diffusion coefficient in the halo and the disk. The parameters of the diffusion equation can be obtained by taking a source distribution proportional to the supernova distribution and measuring the local density and energy spectra of the cosmic rays, which depend on the transport parameters, gas densities and magnetic fields between the source and the Solar system. Such a model is however incomplete. The whole of our knowledge is based on measurements done only at one point on the outskirts of the Galaxy, the solar system, and the assumption that particle spectra and composition are (almost) the same at every point of the Galaxy. γ-rays are able to deliver information directly from distant regions thus complementing our knowledge obtained from cosmic-ray measurements. Some part of the diffuse γ-rays is produced by energetic nucleon interactions with the gas via neutral pion production, another is produced by electrons via inverse Compton scattering and bremsstrahlung. These processes are dominant in different parts of the spectra of γ-rays, therefore, if deciphered the γ-ray spectrum can provide information about the large-scale spectra of nucleonic and leptonic components of cosmic rays.

In addition to diffusion, convective transport modes are expected to play a role for cosmic rays. Supernovae eject hot gas into the interstellar medium which can expand into the halo, presumably driven by the comsic ray pressure from supernova remnants. Such Galactic winds can blow cosmic rays out of the Galaxy, most efficiently at radii with high source density. This mechanism is called convection. It has been proposed as an explanation for the relatively small production of diffuse γ-rays near the sources in comparison to the γ-ray production at larger radii, known as the "soft γ-ray gradient problem".

In fact, a recent analysis of the ROSAT data on X-rays implies wind speeds of up to 760 km/s in the halo. Speeds like this are sufficient to solve the soft-gamma-ray-gradient problem. Unfortunately, the maximum allowed convection speed in isotropic propagation models is restricted to a few tens of km/s, because otherwise cosmic rays leave the Galaxy too fast and the constraints from the relative abundances of cosmic rays and the cosmic clocks cannot be met. Wind velocities as small as this are incompatible with the ROSAT observation.
An additional problem for isotropic diffusion models, is the large bulge-over-disk ratio of the 511 MeV positron annihilation line, as observed by the INTEGRAL satellite. These low energy positrons are thought to originate predominantly from the decay of radioactive nuclei such as ^{56}Co, produced in SNIa explosions in the bulge and in the disk. Predictions of the bulge-over-disk ratio from the expected number of SN explosions are well below one, because of the high rate of SNIa explosions in the disk. However, INTEGRAL found a bulge-over-disk ratio of a few. In a diffusion model without convection MeV positrons hardly propagate, because diffusion is proportional to the velocity and energy of the particle. In this case positrons annihilate close to their sources leading to a small bulge-over-disk ratio. Convective transport in Galactic winds is independent of energy, so low energy positrons in the disk can be convected to the halo, where there are hardly electrons at rest to annihilate with. In this case the escape fraction of positrons from the disk can be sufficient to explain the observed large bulge-over-disk ratio.

About this thesis

In this thesis an anisotropic transport model, which allows for ROSAT compatible convection and still meets all the constraints from primary and secondary cosmic rays, cosmic clocks, γ-rays and the INTEGRAL bulge-over-disk ratio, is introduced. The model features different diffusion coefficients in the halo and in the disk. By increasing the diffusion coefficient towards the halo boundary a smooth transition to free escape is obtained, so the model becomes insensitive to the precise position of the boundary. This is in strong contrast to isotropic propagation models, which require a precise size of the halo for a given diffusion constant. The model has been implemented in the public GALPROP code, the up to now most advanced program for Galactic cosmic ray transport, which solves the cosmic ray transport equation numerically, taking into account the effects of diffusion, Galactic winds, diffusive reacceleration and momentum losses. The code incorporates a

realistic large scale model of the interstellar medium and the radiation fields.
In the framework of this study an extended GALPROP version, which allows for an arbitrary grid spacing and an arbitrary spatial dependence of all transport parameters, has been developed. The launch of the PAMELA and Fermi satellites has significantly increased the amount of accurate data on Galactic cosmic rays in the past year. A further improvement is expected from the AMS-02 instrument onboard the International Space Station, which is scheduled to be launched on a shuttle flight in 2010. With the increasing level of data accuracy, small-scale variations in transport parameters and gas densities can no longer be neglected. The extended GALPROP version developed here allows for the first time to examine the impact of small-scale structures such as the spiral structure of the Milky Way and the so-called Local Bubble, a low density region surrounding the Sun, upon the local cosmic ray fluxes and diffuse γ-rays. In addition the stable products of dark matter annihilation were implemented as a source term in the code. This allows us to examine the impact of a possible contribution from dark matter annihilation upon the fluxes of charged cosmic rays and diffuse γ-rays.

Chapter 2

This chapter reviews the basic observational knowledge about cosmic rays and the implications about the elementary processes of cosmic ray transport and cosmic ray sources that can be derived from there. Starting with a phenomenological diffusion model, the simplest model for cosmic ray transport, the so-called leaky-box model, is introduced and its predictions and limits are discussed. Leaky-box models neglect important processes, such as diffusive reacceleration (2nd order Fermi acceleration) and convection (Galactic winds), which are expected to be relevant for cosmic ray transport. To establish a complete description of all relevant processes the full transport equation is derived from the Boltzmann equation. We explicitly include convection in this equation as a general movement of the background medium.

Chapter 3

The GALPROP code, which solves the full transport equation for cosmic rays numerically, is introduced in Chapter 3 and the Galactic features, such as gas content, supernova distribution, interstellar radiation fields and magnetic field, used in this program are discussed. The GALPROP models feature isotropic and spatially constant diffusion. Such models are able to describe a large number of observations and the GALPROP code has proven to be a useful tool for estimates of the Galactic cosmic ray density. However, these models fail when significant wind velocities as expected from ROSAT are invoked. In addition these models do not allow for spatial variations and possible anisotropies in the diffusion coefficient. The size of the halo, i.e. the region in which cosmic rays are assumed to be bound to the Galaxy via diffusion, is an important parameter of these models and greatly influences the transport parameters required to get a good description of the locally measured cosmic ray fluxes and their relative abundances. Above the halo boundary free escape of comsic rays is assumed. This is an unrealistic scenario, since the scattering rate, which is assumed to be constant throughout the Galaxy, abruptly drops to zero at this boundary. Moreover, in reality high energy cosmic rays are expected to fill a larger volume than low energy cosmic rays, which would mean that the halo size is energy dependent. As a result of the spatial homogeneity of the transport parameters and the gas distribution, the INTEGRAL observation of the large bulge-over-disk ratio in positron annihilation cannot be explained in isotropic transport models.

Chapter 4

We show that the strict constraints on the maximum wind velocity in isotropic models can be loosened if one takes into account that the velocity in the Galactic plane, at the base of the wind, may be non-zero. This way the halo region where convection dominates over diffusion starts closer to the Galactic disk. Convective transport is energy independent, while diffusion increases with energy. In order to explain the exact energy dependence of cosmic ray escape one has to rely on diffusion. Here we show that in the case of non negligible convection in the halo, the diffusion coefficient in the halo has to be larger than the diffusion coefficient in the disk. This way high-energy cosmic rays can diffuse further into the halo by virtue of the increased diffusion coefficient above the Galactic plane and this way the energy dependence of cosmic ray escape is incorporated naturally in the transport parameters. Since the mathematical boundary condition is in a region where convection dominates and the probability for a cosmic ray to return to the disk is small, the model predictions are virtually independent of the size of the transport box. Such anisotropic propagation models (aPM, where the term "anisotropic" refers to the fact that cosmic ray transport perpendicular to the disk is driven by diffusion and convection, while transport in radial direction is only due to diffusion) lead to a significant reduction of the amount of MeV positrons in the disk. This way the absence of positron annihilation from positrons from ^{56}Co decays in the disk can be explained in a natural way.

Chapter 5

In this chapter we examine the uncertainties in the density and spectra of charged cosmic rays in an anisotropic transport model. Such uncertainties arise from variations in the Galactic gas density or the transport parameters. In addition there might be a significant contribution of stable decay products from dark matter annihilation in the local fluxes of cosmic rays and in the diffuse γ-rays.

Variations in the Galactic gas density and the transport parameters are associated to known structures like the Milky Way's spiral arms or the Local Bubble, a low density region of space surrounding the Sun which was created by the explosion of around 14 supernovae more than 10 Myrs ago. The locally measured secondaries are predominantly produced in the local interstellar medium, while the diffuse γ-rays are collected from the entire Galaxy. This means that variations in the local gas density can be probed by comparing, e.g. the local ratio of secondary cosmic rays to primary cosmic rays to the diffuse γ-rays. The anisotropic transport model developed in Chapter 4 offers the possibility to implement such structures at a reasonable level of detail. It is shown that variations in the local gas density have a significant impact on the relative abundances of charged cosmic rays. We discuss how this can be used in order to explain the preliminary Fermi-LAT data on diffuse γ-rays, which show a slightly different normalization than the local cosmic ray fluxes, in a self-consistent way.

Furthermore, we discuss the dark matter interpretation of the EGRET data on diffuse γ-rays in an anisotropic propagation model and compare the dark matter profile derived from the EGRET data to the preliminary Fermi-LAT data on diffuse γ-rays. This dark matter profile consists of a triaxial halo and two concentric rings, with the Sun located in between these rings. We show that an additional contribution from dark matter annihilation improves the predicted flux of diffuse γ-rays, but the additional antiproton component from dark matter annihilation overshoots the data due to the strong contribution from the inner ring of dark matter. With the help of the code developed in the framework of this thesis we can go to a further level of detail and examine the impact of anisotropies of the diffusion coefficient. It is shown that if the diffusion coefficient in vertical direction (for transport away from the Galactic disk) is slightly larger than the diffusion coefficient in radial direction (for transport parallel to the Galactic disk) the antiproton contribution from the inner ring can be greatly reduced. Chapter 5 is closed with a discussion of contemporary indirect dark matter searches from the viewpoint of cosmic ray transport uncertainties in the anisotropic transport model.

Chapter 6

The thesis is concluded with a summary of the results and an outlook to future studies.

Some of the figures in this thesis are originally color figures. An electronic version with color figures can be obtained from: http://digbib.ubka.uni-karlsruhe.de/volltexte/1000015823.

Chapter 2

A Physicist's Guide to the Galaxy

On the morning of August 7, 1912, the Austrian physicist Victor Hess and two companions, Captain W. Hoffory and the meteorological observer W. Wolf, climbed into a balloon gondola for the last of a series of seven flights over Austria, Bohemia and Prussia. During the following six hours Hess carefully recorded the readings of three electroscopes which he used to measure the intensity of radiation. The flight started at Aussig on the river Elbe and as the balloon gained altitude Hess noted a rise in radiation. By noon the group landed at Peiskow, some 50 km from Berlin. In the *Physikalische Zeitschrift* of November 1st that year, Hess wrote "The results of these observations seem best explained by a radiation of great penetrating power entering our atmosphere from above ..." (Hess, 1912). This event was the beginning of cosmic ray astronomy. Twenty four years later Hess shared the Nobel price in physics for his discovery with Carl Anderson for the discovery of the positron. Notably, Anderson discovered the positron by observing cosmic ray tracks in a cloud chamber (Anderson, 1933) and in the following years a number of important discoveries in particle physics have been made in cosmic rays. For example, the muon (Neddermeyer & Anderson, 1937; Street & Stevenson, 1937), the pion (Perkins, 1947) and the kaon (Rochester & Butler, 1947), among others, were first observed in experiments studying cosmic rays. The interplay between the study of cosmic ray physics and conventional particle physics has created the field of astroparticle physics. On the other hand, particle physics plays a vital role in the understanding of cosmic-ray production and propagation mechanisms. Even today the highest energies of single particles are not produced in particle accelerators on Earth but in Galactic and intergalactic space. The components of cosmic rays can be divided into charged and neutral particles. Charged particles are influenced by magnetic and radiation fields so they do not point back to their sources as neutral particles do. Neutral particles, such as high energetic photons or neutrinos, are partially produced directly in the cosmic ray source regions, which are then visible to us e.g. in γ-rays and partially produced by the subsequent interactions of charged cosmic-rays in the interstellar medium. While there is agreement on the Galactic production sites and acceleration mechanisms, for the highest energies the sources of cosmic rays and in particular the acceleration mechanisms are not completely understood. A review on cosmic rays and their propagation and acceleration can be found in standard textbooks, e.g. Gaisser (1990); Ginzburg et al. (1990); Longair (1992); Schlickeiser (2002).

In this chapter the basics of cosmic ray physics are introduced. We will begin by briefly reviewing the current observational knowledge about cosmic rays, their sources and their propagation in the Milky Way. In this thesis we are only interested in Galactic cosmic rays, which brings us into the fortunate position of dealing with a topic on which basic agreement has been reached, although the details are still subject to discussion. Having established the observational constraints we then discuss the basics of cosmic ray propagation which are required to derive the transport equation for cosmic rays. We will briefly discuss the limits of the isotropic diffusion approximation and then turn to the more general anisotropic diffusion-convection equation. This equation will form the basis for all subsequent discussion of cosmic ray transport.

2.1 Observational Constraints

In many fields of physics data can be retrieved under controlled conditions in laboratory experiments. The system of interest can be embedded in an experimental setup, which allows us to control the processes that we examine. For the field of astrophysics this is not possible. Compared to other fields of physics the observational knowledge in astrophysics and astronomy is limited in the sense that a significant amount of interpretation of the raw data is required. Unavoidably, this leads to observational artifacts, which have to be considered whenever dealing with astronomical data. This is mainly due to the fact that astronomical measurements, such as gas densities or field strengths suffer from large model uncertainties due to the fact that only column densities can be observed. For cosmic ray physics the situation is similar. Being limited to a single point in the Galaxy the density and the energy spectrum of cosmic rays is only known for the position of the Sun. Any information about the cosmic rays spectra in other parts of the Galaxy has to be derived indirectly from diffuse γ-rays or synchrotron radiation which are produced by the interaction of cosmic rays with the interstellar medium (ISM), the Galactic magnetic field or the interstellar radiation field (ISRF).

In this section we review observational constraints for cosmic ray (CR) transport models. Starting with the basic knowledge about the CR spectra observed at Earth we will turn to promising sources of Galactic cosmic rays and then discuss the immediate consequences for CR transport.

2.1.1 The Composition and the Spectrum of Cosmic Rays

While at low energies the CR spectrum mainly consists of protons and light elements, the fraction of heavier elements increases with increasing energy significantly. At around 100 GeV/nucleon protons make up about 56% of the cosmic rays, helium 24% and heavier elements 20%. At 1 PeV/nucleon the spectrum consists of about 15% protons, 33% helium and 52% heavier elements (Nilsen, 1998).
The left side of figure 2.1 shows the CR energy spectrum. The primary cosmic-ray particles extend over at least 13 decades in energy with a corresponding decline in intensity of over 32 decades. Below 10^{16} eV the spectrum is remarkably featureless with little deviations from a power law with spectral index -2.7, above this value the energy dependence becomes E^{-3}. The turning point, known as the knee, has a flux of about $1\frac{1}{m^2 \cdot yr}$. A second change in the gradient of the spectrum occurs at 10^{19} eV, known as the ankle, where the spectrum becomes less steep once again.
The origin of the knee is still subject to discussion. It could be caused by different effects, such as different source populations (Galactic supernovae (SNs) for low energies and active galactic nuclei (AGNs) for high energies), different energy loss mechanisms below and above the knee or a change in the elemental composition (Asakimori et al., 1998). As an alternative new interaction characteristics owing to new particle physics at energies above 1 TeV/nucleon (Asakimori et al., 1998) or an observational bias related to the change in the experimental technique from direct particle-by-particle balloon and spacecraft experiments below $\sim 10^{14}$ eV to indirect ground based air shower measurements above 10^{15} eV (Asakimori et al., 1998) have been suggested. Below 10^{19} eV it is not possible to track particles back to their sources, even if the arrival direction at Earth is known. The trajectories of these particles are completely randomized by the galactic magnetic field, since, even at these energies, the gyro-radius in the galactic magnetic field is smaller than the size of the

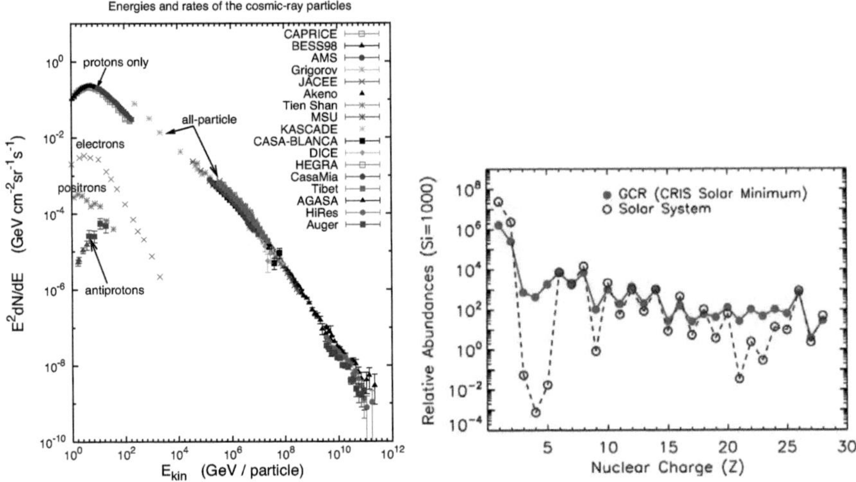

Figure 2.1: Left: Compilation of measurements of the energy spectrum of charged cosmic rays. The observations can be described by a power-law with spectral breaks at 4 PeV, referred to as the knee, a second knee at 400 PeV and the ankle at 1 EeV (Gaisser, 2007). **Right:** Comparison of CR (here labelled GCR for Galactic cosmic rays) abundances (filled circles) to the Solar system abundances (open circles) from George et al. (2009). The CRIS Solar minimum results reported in George et al. (2009) are used for the $z \geq 5$ GCR abundances. For $Z < 5$, the GCR data come from Wang et al. (2002) and de Nolfo et al. (2006). The Solar system abundances are taken from Lodders (2003) and represent the abundances of elements in the proto-Solar nebula.

Galaxy. Assuming a magnetic field of $1\,\mu$G and a Galactic radius of 50.000 Ly the gyro-radius of a proton becomes comparable to the radius of the Milky Way at energies above 10^{19} eV. For particles with energies larger than 10^{19} eV the gyro-radius exceeds the size of the Galaxy and these particles are believed to be of extragalactic origin. No phenomenon in the neighborhood of our Galaxy can account for CRs with energies up to 10^{19} eV, yet their sources may not lie much further away, because otherwise the Greisen-Zatsepin-Ku'zmin (GZK)-cutoff needs to be taken into account. This cutoff is due to CRs interacting with photons from the intergalactic radiation field: Space is filled with the cosmic microwave background (CMB) radiation, a relic of the epoch of recombination when the first hydrogen atoms were formed and the universe became transparent to photons. There are about 10^9

of these photons in a cubic meter of space, yet normally a CR will be oblivious to their presence. This changes, however, when a CR has so much energy that the CMB photon's energy is sufficient to cause the Δ-excitation:

$$p + \gamma_{CMB} \longrightarrow \Delta^+ + X \longrightarrow \begin{cases} n + \pi^+ \\ p + \pi^0 \end{cases} \quad \text{for } E_p - p_p \cdot cos\theta \geq \frac{m_p m_\pi}{q}, \tag{2.1}$$

where E_p is the energy of the proton in the center-of-mass system, p_p is the absolute value of the proton's momentum in the center-of-mass system, θ is the angle under which the proton hits the photon, q the absolute value of the photons momentum in the center-of-mass system, m_p the proton mass and m_π the pion mass. The universe becomes opaque for CR protons when the Δ-excitation becomes energetically allowed at energies larger than $5 \cdot 10^{19}$ eV. The excited state then decays by the two shown channels. The resulting proton will be of lower energy, the resulting cutoff in the proton spectrum is called GZK cutoff. A $5 \cdot 10^{19}$ eV proton is expected to be reduced to an energy below the cutoff over a distance of 50 Mpc[1] . Particles with energies above the cutoff have been detected nonetheless, despite the lack of known sources within range.

Recently, the Pierre Auger collaboration has reported a tentative correlation between the arrival directions of cosmic rays above $6 \cdot 10^{19}$ eV and the position of active galactic nuclei within ~75 Mpc (Pierre Auger Collaboration, 2007), but the small number of events above 10^{20} eV, corresponding to macroscopic energies of a few Joules, confirms the GZK cutoff.

In this work the energy range of interest lies below 10^{15} eV. In this energy range CRs are mainly of Galactic origin, with only a negligible fraction of extragalactic CRs. Below 1 GeV the flux drops dramatically below the extended power law as can be seen from the left side of Fig. 2.1. This is an artifact, because at low enough energy the Solar wind and its associated magnetic field is able to prevent the propagation of charged particles into the heliopause.

At the position of the Earth the relative abundances of the more common elements (including C, O, Ne, Mg, Si, Fe, Ni) are remarkably similar to the relative abundances of these elements in the Solar system. The right side of figure 2.1 shows the CR and local abundances for the elements from H to Zn as seen by CRIS (George et al., 2009).

[1] 1 pc$\approx 3.086 \cdot 10^{16}$ m. For comparison: The diameter of the Milky Way is about 30 kpc; the nearest spiral galaxy, the Andromeda Galaxy (M31) is 770 kpc away from us; clusters of galaxies, which contain 500-1000 galaxies have typical diameters of 2-10 Mpc; superclusters (i.e. clusters of galaxy clusters), which are among the largest structures in the universe, have been observed with diameters between 10 and 100 Mpc; the particle horizon (the boundary of the observable universe) has a radius of about 14 Gpc.

Nuclei heavier than ^7Li have to be produced by fusion of lighter nuclei in stars, because there are no stable elements with $A = 5$ or $A = 8$ which could serve as intermediate steps during primordial nucleosynthesis. On the other hand, many elements that are rare in the Solar system (such as Li, Be, B, F, Sc, Ti, V) occur with much higher abundances in the arriving CRs. Notably, these are elements which are essentially absent in stellar nucleosynthesis.

This discrepancy between the local elemental composition and the elemental composition of CRs has been well known for many years and is an important key to understanding CR transport. On the one hand the similarity between the CR and Solar system abundances of the more common elements implies that the composition of the CR source material, which was accelerated on the order of 10^7 years ago, is very similar to that of the nebula that formed the Solar system 4.6 x 10^9 years ago. On the other hand the fact that the less common elements, which are not produced in suns, are much more abundant in the arriving CRs than in the Solar system, can be understood quantitatively as the result of nuclear interactions of abundant cosmic ray elements with interstellar gas. As an example, interactions of C, N, O result in fragments of lighter elements, ^3Li, ^4Be and ^5B.

Another noteworthy difference is that protons are much more abundant in the Solar system than in CRs. The likely cause of this discrepancy is that hydrogen is comparably hard to ionize and thus only small amounts enter into the acceleration process for CRs.

2.1.2 Sources of Galactic Cosmic Rays

In this work we are interested in an energy range far below the knee, i.e. in energies of $10^{-1} - 10^2$ GeV. Sources of CRs at these energies are almost exclusively Galactic. Particle acceleration up to GeV energies has been observed in solar flares, for acceleration up to TeV energies a different mechanism is required.

As a rough approximation the local CR energy density, which is roughly $\rho_E \approx 1\,\mathrm{eV/cm^3}$, can be assumed to be representative for the whole Galaxy. Taking the timescale for diffusion out of the Galaxy to be $\tau \sim 2 \cdot 10^7$ yrs and assuming a Galactic radius of 15 kpc and a disk height of 800 pc we can estimate the power required to keep up the CR energy density to be

$$P_{\mathrm{CR}} = \frac{V \rho_E}{\tau} \sim 4 \cdot 10^{33}\,\mathrm{J/s}. \tag{2.2}$$

Among the known Galactic objects, supernovae are the best candidate for this power. With a mean rate of one supernova per 30 years and a typical energy release of 10^{44} J per

supernova, the total power released by supernovae is of the order 10^{35} J/s. The efficiency with which the energy is transported to CRs is unknown, but from Eq. 2.2 it is clear that only a few percent would suffice. The left side of Fig. 2.2 shows Tycho's SNR as seen by CHANDRA in order to give an impression of these objects.
The energy transport in supernova remnants is believed to be dominated by the so-called first order Fermi acceleration. In this acceleration mechanism nuclei and electrons ejected from the dying star scatter on magnetic turbulences carried along with the expanding supernova shell. For each passage through the shock front the energy gains are proportional to v/c, where v is the velocity of the shock front. Since for each cycle there is a certain probability for the particle to be lost from the acceleration region, this process naturally leads to the observed power law in the CR spectrum. We will discuss the Fermi acceleration mechanism in Section 2.3.2 in more detail. Recently, the idea that CRs are accelerated to the TeV rage has been supported by observational evidence. The right side of Fig. 2.2 shows the supernova remnant RX 1713.7-3946 as seen by the HESS telescope in γ rays. The observed γ-ray spectrum follows a power law with spectral index of about 2 and extends up to roughly 10 TeV. This implies the presence of protons accelerated to even higher energies than that, producing the observed γ-rays in collisions with matter present in the vicinity of the supernova remnant.
The solar nuclear fusion chain stops with ^{56}Ni, which is unable to release energy by fusion, but does produce ^{56}Fe through radioactive decay. Heavier elements are produced through endoterm reactions in SNs, but their abundances in CRs are negligible. In this work we only consider nuclei up to Ni.

2.1.3 Propagation of Cosmic Rays in the Galaxy

The local CR density, the energy spectrum and the relative abundances of CRs are the only direct information about CRs we can obtain. Together with the estimates of the CR column density which we can deduce from the diffuse γ-rays this observation forms the basis for any model for CR transport. The featureless spectrum below the knee is believed to indicate that CR production and propagation is governed by the same mechanism in this energy range.
The initial spectra of CRs are modified by a variety of physical processes on their way through the Galaxy. Likewise, their initial composition is changed by hadronic interactions of protons and nuclei with the interstellar matter. Purely secondary CRs, such as boron (B), beryllium (Be), antiprotons and positrons, which do not originate from SNRs, are

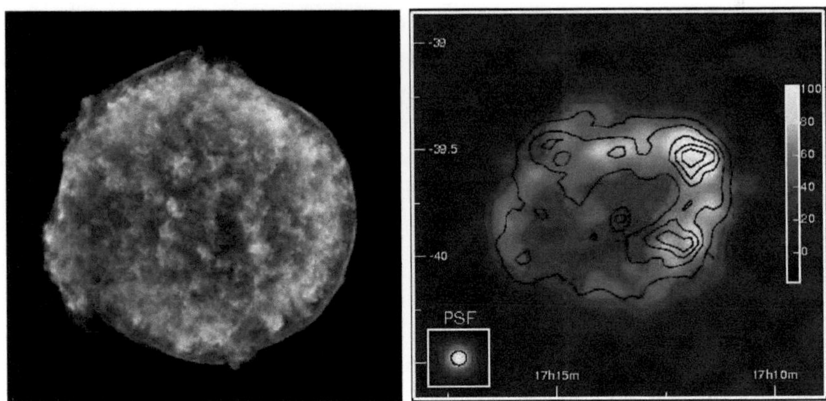

Figure 2.2: Left: 2005 CHANDRA image of Tycho's SNR, from Chandra.nasa.gov. The outer thin surface is synchrotron radiation from highly relativistic electrons accelerated at the outer shock. Behind this is a highly turbulent region, presumably formed by Rayleigh-Taylor instability at the contact discontinuity. Acknowledgement to NASA/CXC/SAO. **Right:** Acceptance corrected γ-ray excess image from supernova remnant RX 1713.7-3946 as measured by HESS (Aharonian et al., 2007). Shown is a combined image from the 2004 and 2005 data. The contour lines trace the x-ray emission.

produced. A fraction of the radioactive isotopes, like ^{10}Be, decays in flight before reaching the detector. Hadronic interaction also produce γ-rays by π^0-production and consequent decay, electrons and positrons lose energy very efficiently by synchrotron radiation in the Galactic magnetic field, bremsstrahlung in the interstellar material and inverse Compton scattering on photons of the interstellar radiation field (emission from stars, dust and CMB). Since the Galaxy is basically transparent for γ-rays, the photons produced in CR interactions will follow straight lines and provide information about the line of sight integral of the CR and the gas or field density required for their production.

For CRs the situation is more complicated, because their paths are bend by the Galactic magnetic field. Since the production and fragmentation cross sections are known from fixed target experiments at Earth one can estimate the amount of matter CRs traversed on their way to the detector from the relative abundance of secondary CRs. More specifically, one can deduce the grammage from the ratio of secondary to primary CRs. The grammage χ

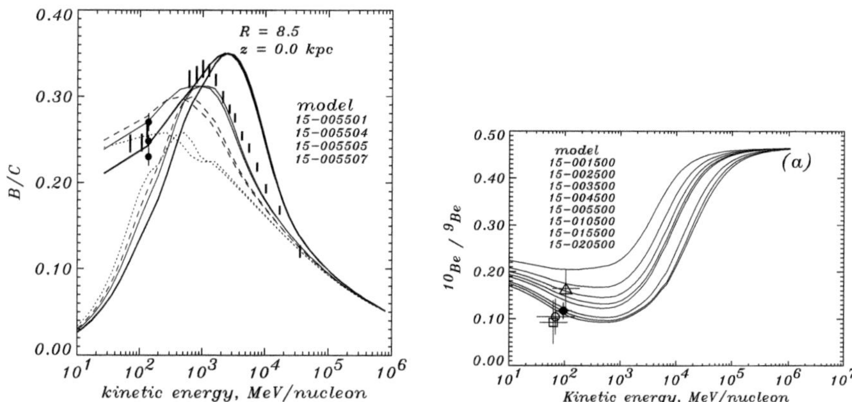

Figure 2.3: B/C and $^{10}Be/^{9}Be$ ratios for diffusive reacceleration models from Strong & Moskalenko (1998). **Left** B/C ratio for diffusive reacceleration models with a halo height of $z_h = 5$ kpc, v_α=0 (dotted), 15 (dashed), 20 (thin solid), 30 kms^{-1} (thick solid). In each case the interstellar ratio and the ratio modulated to 500 MV is shown. Data: vertical bars: HEAO-3, Voyager Webber et al. (1996), filled circles: Ulysses DuVernois et al. (1996): Φ= 600, 840, 1080 MV). **Right** $^{10}Be/^{9}Be$ ratio for diffusive reacceleration models as function of energy for (from top to bottom) z_h =1,2,3,4,5,10,15 and 20 kpc, the data points are from Lukasiak (1994) (Voyager-1,2: square, IMP-7/8: open circle, ISEE-3: triangle) and Connell et al. (1998) (Ulysses): filled circle.

is defined as the gas column density along a CRs path

$$\chi = \int n_H c \tau_{esc}, \tag{2.3}$$

with n_H the averaged gas density and τ_{esc} the time CRs spend in the Galaxy. The boron-to-carbon (B/C) ratio shown on the left side of Fig. 2.3 is an example for such a measurement, since B is predominantly produced from C, N, O. Figure 2.3 also shows the prediction of different diffusive reacceleration models, which will be discussed later in this work. At this point it is only important for us to note that an increase in halo height increases the secondary to primary ratios, because CR spend longer times in the Galaxy. Similarly, the relative abundances of radioactive instable isotopes can be used as "cosmic clocks" in the sense that the local age of CRs can be estimated from the ratio of radioactive instable to radioactive stable isotopes. An example for such a "cosmic clock" is the ratio of ^{10}Be over

^9Be. This is shown on the right side of Fig.2.3. Again, the prediction of different diffusive reacceleration model is shown together with the data. With increasing halo height the time CRs spend in the Galaxy increases and the fraction of "surviving" ^{10}Be decreases.
A comparison of these two measurements to the average gas density in our Galaxy leads to the conclusion that CRs have to spend most of their time in regions with low density, such as the halo. Thus there has to be a process which efficiently confines CRs to the Galaxy and allows them to return from the halo to the Galactic disk. A pure propagation of the CRs guiding center along magnetic field lines therefore can be excluded, because this process does not offer the possibility to return from the halo. CRs therefore are assumed to scatter on turbulences in the magnetic field. If this scattering is resonant, it can be modelled by a diffusion equation with a diffusion coefficient depending on the particle's rigidity. The origin of the magnetic turbulences is unknown. For small energies CRs can scatter on turbulences that the CRs themselves create (Cesarsky, 1980). These are the so-called Alfvén waves. Alfvén waves grow in amplitude as the result of the scattering of high energy particles by the magnetic field in the waves, until the streaming velocity of the high energy particles is reduced to the Alfvén velocity $v_\alpha \approx B_0/\sqrt{\mu_0\rho}$, where ρ is the mass density of the fully ionized plasma. However, this is only effective if the waves are not damped before they have time to grow to significant amplitude. The presence of neutral particles in the interstellar plasma can readily abstract energy from the Alfvén waves by neutral-ion collisions in a time that is short compared to the growth time. The significance of the neutral particles is that they can remove kinetic energy from the waves, whereas ionized particles are simply constrained to oscillate with the waves. Consequently, in regions with a large density of neutral material, wave damping is expected. The typical timescale for the growth of the waves is a strong function of particle energy. For low energies ($\sim$ 3 GeV), the high energy particles are sufficiently numerous, so that the growth rate exceeds the damping rate. For energies above $10^2 - 10^3$GeV, the damping rates become too large and this is no longer possible (Cesarsky, 1980). At least for these energies an external source of magnetic turbulence, such as rotating stellar winds, rotating white dwarfs, magnetic stars or molecular clouds have to be considered. It should be noted that the magnetic turbulences are not a directly observed quantity, but one that is inferred from the large confinement volume and large confinement time of CRs.

Locally, CR diffusion is highly anisotropic and occurs along the magnetic field lines. However, if the magnetic field has no preferred direction, i.e. if the turbulent small scale component ($\sim 100\ pc$) is larger than the unperturbed component, CR diffusion is assumed

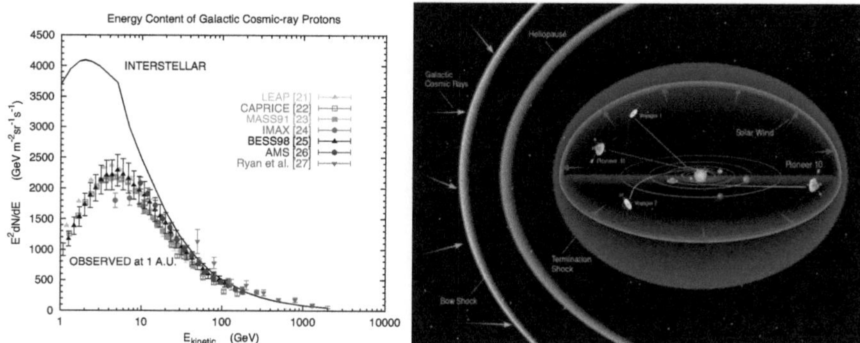

Figure 2.4: Left: Summary of direct measurements proton spectra made with detectors on balloons and in space. The line labelled "interstellar" indicates the possible local interstellar proton flux. From Gaisser (2001). **Right:** Schematic view of the heliosphere. Source: Wikipedia.

to be isotropized by pitch angle scattering on the magnetic turbulences.

Solar and geomagnetic modulation

The left side of Fig. 2.4 shows the proton spectrum for energies between 1 and 10^3 GeV. In this energy range a simple power law approximation does not suffice to describe the data. For energies below 10 GeV the spectrum of locally measured CRs differs significantly from the local interstellar spectrum (LIS). This is due to the interaction of CRs with the Solar wind (Longair, 1992). As the wind begins to collide with the interstellar medium, it slows down before finally ceasing altogether. The point where the Solar wind slows down is is called the termination shock; the point where the interstellar medium and Solar wind pressures balance is called the heliopause; the point where the interstellar medium, travelling in the opposite direction, slows down as it collides with the heliosphere is the bow shock. The right side of Fig. 2.4 shows a diagram of the features of the heliosphere. The strength of the Solar wind varies depending on the Sun's activity. The first hints at this effect came from observations of an anticorrelation between neutron monitor counts and the sunspot number, the latter being an indicator of the level of Solar activity (Stanev, 2004). The wind originates from the corona of the Sun and consists mainly of protons with a typical flux of $1.5 \cdot 10^{12} \mathrm{m}^{-2}\mathrm{s}^{-1}$ at a typical kinetic energy of 500 eV. This corresponds to a velocity of 350 km/s and a temperature of 10^6 K. The Sun's magnetic field is frozen into the Solar wind plasma, leading to the creation of the so-called Archimedes spiral due to the Sun's rotation. This is shown on the left side of Fig. 2.5. CRs scatter on the magnetic turbulences in this plasma. Due to the limited size of the system under consideration, Solar modulation can only be efficient for smaller energies typically below 10 GeV/nucleon. Solar modulation can be modelled by CR diffusion in a turbulent magnetic field, convection by the outward motion of the Solar wind, and adiabatic energy losses in this flow (Gleeson & Axford, 1968). We will discuss this in some detail in Section 2.4.7.

A second modulation occurs in the so-called van-Allen belts of the Earth's magnetic field. Close to the Earth the magnetic field can be approximated by a dipole field. The high field densities at the poles form magnetic mirrors and low energetic CRs can be reflected between the mirror points multiple times and eventually lose their energy. The right side of Fig. 2.5 shows an artist's impression of this process. The minimal rigidity that a particle must have in order to reach Earth is called the cutoff rigidity R_S. For the case of a dipole, the cutoff rigidity for vertically approaching particles is given by (Longair, 1992):

$$R_S \geq 14.9\,\mathrm{GV} \cdot \cos^4 \lambda. \tag{2.4}$$

Here λ is the latitude of the observer with respect to the equatorial plane of the dipole. Near the magnetic poles the cutoff becomes minimal. Equation 2.4 only holds for vertically approaching particles. In reality the flux of CRs at the position of the Earth is almost isotropic. The cutoff also depends on the azimuthal angle, leading to the *east-west-effect*: For the same zenith angle the cutoff for particles with positive charge approaching from the east is higher than for positive particles approaching from the west (and vice versa for particles with negative charge) (Stanev, 2004).

Since the Earth's magnetic field is far from being a perfect dipole and in addition suffers time dependent deformations due to distortions caused by the Solar wind an accurate determination of the geomagnetic cutoff requires a detailed model of the geomagnetic field. This can be done by backtracing individual particles with a given position, time, and rigidity through the magnetic field by integrating the equation of motion, to see if the particle reaches outer space.

The geomagnetic cutoff leads to an additional modulation of the CR spectrum $\Phi(R)$. It

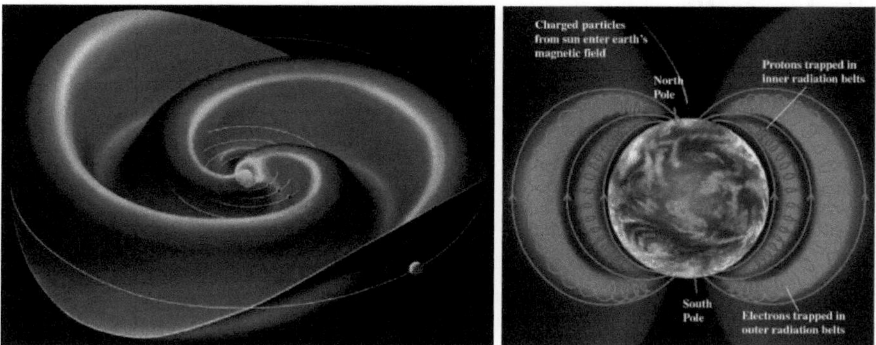

Figure 2.5: Left: The heliospheric current sheet out to the orbit of Jupiter. Source: Wikipedia. **Right:** An artist's impression of the Van-Allen-belts. Source: Wikipedia.

can be described as (Mizuno, 2001)

$$\Phi^{\text{mod,geo}}(R) = \Phi(R) \cdot \frac{1}{1 + \left(\frac{R}{R_c}\right)^{-\gamma_c}}, \tag{2.5}$$

where R_c is a cutoff rigidity. γ_c describes the steepness of the modulation and is typically of the order of unity. Since in this study we use data which have been taken during different phases of the Solar cycle, we treat the Solar modulation strength as a free parameter. Consequently, the additional geomagnetic modulation vanishes in the uncertainties of the Solar modulation strength, so that in the following we can neglect this modulation.

2.2 Detection Techniques for Cosmic Rays

Depending on the energy of interest CR experiments are either indirect ground based detectors, which can be both, pointed instruments (with a small field of view) for γ-rays and instruments with a wide field of view for both charged CRs and γ-rays, or high altitude direct detectors.

At sea level the thickness of Earth's atmosphere amounts to twenty radiation lengths or eight hadronic interaction lengths. CRs below 10^{14} eV are absorbed before reaching the surface of the Earth, but for these energies the CR flux is sufficient for a direct measurement. In this case , calorimeters, emulsion chambers, scintillators, tracking devices, Cerenkov counters or transition radiation detectors, are combined in a suitable way to

measure mass, momentum, energy or charge sign of a particle. Such detectors are usually flown on long balloon flights or placed on satellites. This limits the weight of the detector and thus the maximum CR energy to be detected. The PAMELA and AMS-02 detectors are examples of this category. It should be noted that even in the case of high altitude experiments an additional modification of the CR spectra in the thin layer of atmosphere above the detector has to be taken into account.
For energies larger than 10^{14} eV the low CR flux requires large detectors for acceptable counting rates. For these energies the Earth's atmosphere is used as a calorimeter. A CR will interact with a nucleus in the atmosphere (primarily oxygen and nitrogen). The energy of the primary can be derived from the size and shape of the resulting shower at observation level. Extensive air shower arrays (EAS) use e.g. scintillators or water Cerenkov detectors spread over a large area to detect the shower particles. The observation altitude determines the energy threshold for primary CR detection. The quality of shower reconstruction depends on the declination of the primary CR, but generally this type of detector has a very wide field of view. EAS are capable of continuously monitoring the complete overhead sky, thus leading to very large statistics. Detectors of this type can also be used to reconstruct electromagnetic showers as initiated by gamma rays. With sufficient angular resolution EAS can be operated as wide field of view gamma ray telescopes and used for the detection of transient phenomena such as gamma ray bursts or the observation of point sources such as pulsar nebulae. Apart from that, Cerenkov radiation emitted by the shower particles or the fluorescence light can be observed. The latter is emitted after the excitation of nitrogen molecules in the air by passing shower particles, and with wavelengths in the range from 300 to 400 nm. The Pierre Auger observatory in Argentina employs a hybrid detection technique (Abraham et al., 2004) combining both Cherenkov and fluorescence detectors at the same site thus allowing cross-calibration and reduction of systematic effects that may be peculiar to each technique.
In this work we are interested in energies below 10^6 GeV. Most of the results used in this work originate from balloon borne or space probes.

2.3 Basics of Cosmic Ray Propagation

Using the observational knowledge about Galactic CRs as discussed in the last section, we can now examine the detailed processes that CRs undergo in the ISM and ultimately build a model for CR transport.

We have already established in Section 2.1.3 that the escape time of CRs, as measured by radioactive isotopes, is too large to allow for free propagation along the large scale magnetic field. In addition, the grammage, as inferred from the secondary to primary ratio, is too small to allow for CRs to spend all their time in the Galactic disk. Consequently CRs have to be confined to the Galaxy for a considerable amount of time and in addition they have to spend most of their time in the Galaxy in regions with low gas density, like the Galactic halo. This is possible if CRs efficiently scatter on magnetic turbulences, possibly generated by the CR plasma itself. Once CRs leave the source regions they will begin to scatter on these magnetic turbulences, they will start to loose energy by virtue of interactions with the ISM, the Galactic magnetic field or the interstellar radiation field. In many cases these energy losses lead to the production of γ-rays or synchrotron radiation, which can be used as a tracer of the Galactic CR distribution, since the Galaxy is essentially transparent for γ-rays and synchrotron radiation above a few hundred MHz. On the other hand diffusive reacceleration, i.e. scattering on moving magnetic turbulences, can increase the energy of CRs.
In this section we will review the main mechanisms for energy losses and gains and finally invoke some of these processes in simple diffusion models for CR transport.

2.3.1 Energy Losses and γ-Ray Production

Electrons and Positrons Depending on the energy of interest energy losses for electrons and positrons are dominated by bremsstrahlung, Compton and synchrotron losses or ionization and Coulomb losses. From these five processes the following can be detected at Earth by virtue of the γ emission:

- **bremsstrahlung,** i.e. interaction of high energetic electrons with interstellar gas

$$e + p \longrightarrow e' + p' + \gamma \tag{2.6}$$

- **inverse Compton scattering,** i.e scattering of high energetic electrons and low energetic photons of the interstellar radiation field (emission from stars and dust and the CMB)

$$e + \gamma \longrightarrow e' + \gamma' \tag{2.7}$$

- and **synchrotron radiation.**

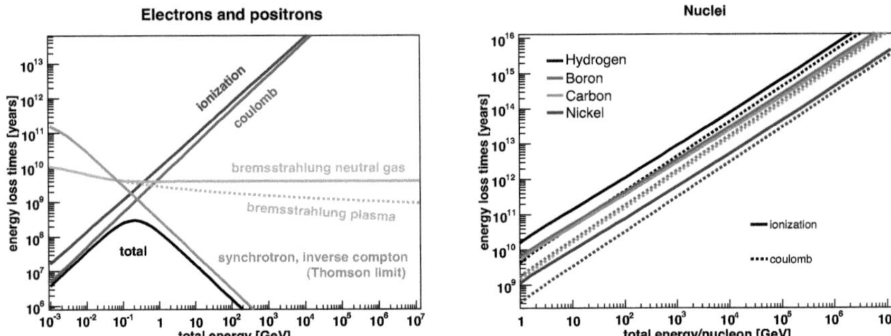

Figure 2.6: Estimate of the time in which CRs loose their kinetic energy. The energy loss times for electrons **(left)** and protons **(right)** are calculated in the Thomson limit, i.e. energy density of photons and magnetic field are identical (=1eV/cm^3) and thus IC and synchrotron losses are identical. The loss times refer to equal gas densities of the neutral and ionized component of the ISM ($n_H = n_{HII} = 0.01/\text{cm}^3$)) and a He to H ratio of 0 (no He). These gas and energy densities are the approximately average values seen by CRs on their way to Earth.

Given that the magnetic field, the interstellar radiation field and the gas densities are sufficiently well known the photons produced in these processes can be used as a tracer of the average Galactic electron and positron distribution. The left side of Fig. 2.6 shows the energy loss times for electrons and positrons. For low energies Coulomb and ionization losses dominate, leading to total energy loss times between $4 \cdot 10^6$ yrs and $3 \cdot 10^8$ years between 10^{-3} and 10^{-1} GeV. For energies larger than a few hundred GeV synchrotron and IC losses become the dominant loss mode and the total energy loss time drops rapidly to 10^6 yrs at $2 \cdot 10^2$ GeV. For energies larger than 10 GeV the electron loss times become comparable to the CR escape time of 10^7 yrs, but even for lower energies the local electron and positron spectra can be significantly modulated by these losses. Details on bremsstrahlung, Compton and synchrotron, ionization and Coulomb losses can be found in Appendices A.1 through A.5.

Protons and Nuclei For protons and nuclei ionization and Coulomb losses are the dominant energy loss mode. The right side of Fig.2.6 shows the energy loss times for different nuclei due to these processes. For all nuclei under consideration the energy loss times exceed the time CRs spend in the Galaxy, which implies that the spectral shapes of protons and nuclei will be practically the same everywhere in the Galaxy. A detailed discussion of proton energy losses due to ionization and Coulomb scattering can be found in Appendices A.4 and A.5.

In addition to ionization and Coulomb losses collisions of heavy nuclei with the hydrogen or helium of the ISM can lead to inelastic scattering with the result that the parent nucleus is destroyed and new CR secondaries are created:

$$n_1 + (p,\ He) \rightarrow n_2 + X, \tag{2.8}$$

here n_1 and n_2 denote the primary and secondary nucleus, respectively. The destruction of primary CRs, given by the total cross section, is called *fragmentation*, while the production of secondary CRs, given by the branching ratio of each channel, is called *spallation*. Spallation and fragmentation are discussed in Appendix A.6 in detail.

One important case to note is the production of neutral π mesons via proton-proton interactions. The consecutive decay of the π^0 mesons leads to the production of 2 photons which again are detectable at Earth:

$$p + p \longrightarrow \pi^0 + X \longrightarrow \gamma\gamma + X. \tag{2.9}$$

Photons produced in this process constitute the main contribution to the diffuse γ-ray flux from the Milky Way in the intermediate GeV range and even dominate the diffuse emission at large latitudes. The spectral shape of these γ-rays is essentially determined by the spectral shape of the proton spectrum along a line of sight.

In addition to fragmentation CRs undergo radioactive decays. Radioactive instable isotopes are created both by fragmentation of heavier nuclei (e.g. ^{10}Be is created from $B,\ C,\ N,\ O$) and directly in CR sources (e.g. ^{26}Al). As we have seen the importance of radioactive unstable elements lies in the possibility of measuring the local average age of CRs. For this purpose isotopes with lifetimes close to the CR escape time, like ^{10}Be are of special interest. In Appendix A.7 we discuss radioactive decays in the framework of CR transport in more detail.

2.3.2 Energy Gains

The initial cosmic ray acceleration in the source regions and the consecutive reacceleration in the ISM is still a subject of intensive studies. At present the scientific community is mostly focused on ultra high energy cosmic rays with energies around 10^{19}eV. The

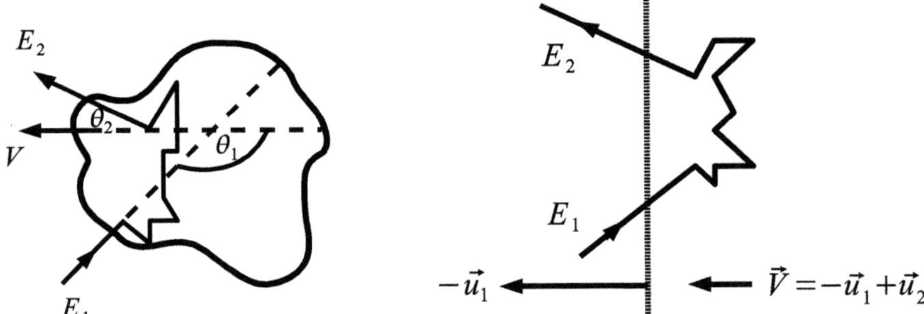

Figure 2.7: Left: Acceleration by a moving, partially ionized gas cloud. **Right:** Acceleration at a plane shock.

acceleration mechanisms in this energy range are still subject to debate, since the theory is not completely established. For Galactic CRs the energies of interest will never exceed the TeV range. For these energies there is good agreement on the mechanisms of CR acceleration and reacceleration.

Although a wide variety of detailed models exists there are two basic acceleration mechanisms, the so-called first order Fermi acceleration and the so-called second order Fermi acceleration, which will be introduced in this section. First order Fermi acceleration is a very efficient process, but it requires the presence of strong shocks. This process is responsible for the initial acceleration of CRs in the source region and may also play a role in the Galactic halo where the expanding SN shells constitute a shock front. Second order Fermi acceleration is less efficient, but it can occur almost everywhere in the ISM, since it only requires randomly moving magnetized clouds.

In the presence of moving magnetic traps a third acceleration mechanism is possible, which is somewhat more efficient than second order Fermi acceleration. Although it is not clear whether or not trap acceleration is present in the ISM we will discuss this mechanism briefly.

Since all acceleration processes in the ISM are magnetic in nature, only charged particles can be accelerated.

Second Order Fermi Acceleration

The explanation of an effective mechanism for particle acceleration suggested by Enrico Fermi in 1949 (Fermi, 1949) was one of the milestones in understanding CR transport.

The original idea was that slowly moving magnetic clouds, about 10-100 times more dense than the interstellar medium and endowed with an enhanced "frozen-in" magnetic field, are responsible for the reacceleration of cosmic rays. These many light years wide clouds are believed to occupy several percent of the ISM. When a charged particle enters such a cloud, it scatters on the random irregularities of the magnetic field that change its momentum with a resulting gain or loss of energy, depending on whether particle and cloud are moving in the opposite or the same direction, respectively.
There are two types of elastic scatterings that end with a reflection of the particle :

- reflection on a "magnetic mirror", i.e the particle is confined in a field flux tube with the field lines focussed towards a single point;
- reflection on a magnetic turbulence, i.e. the particle follows a field line that is bent.

The magnetic irregularities of the field are random in nature. Consequently the scattering inside the cloud can be considered as a random walk. Consider a relativistic particle with energy E_1 in the laboratory frame that enters a slowly moving magnetic cloud as shown in Fig. 2.7. A Lorentz transformation to the rest frame of the moving cloud gives

$$E_1' = \gamma E_1(1 - \beta \cos\theta_1), \tag{2.10}$$

where here and in the following, prime stands for the rest frame of the cloud. θ_1 is the angle between the momentum of the particle and the cloud and β and γ refer to the velocity of the cloud. Since we only consider elastic scatterings due to the motion of the magnetic field, the energy of the particle in the clouds rest frame just before it escapes is $E_2' = E_1'$. Transforming back to the laboratory frame, we get the energy of the particle after its encounter with the cloud,

$$E_2 = \gamma E_2'(1 + \beta \cos\theta_2'). \tag{2.11}$$

Notice, that the sign of β is opposite to 2.10. With 2.10 and 2.11 we get

$$E_2 = \gamma^2 E_1(1 - \beta \cos\theta_1)(1 + \beta \cos\theta_2'), \tag{2.12}$$

and the energy change for the particular encounter characterized by θ_1 and θ_2 is

$$\frac{\Delta E}{E} := \frac{E_2 - E_1}{E_1} = \frac{1 - \beta\cos\theta_1 + \beta\cos\theta_2' - \beta^2\cos\theta_1\cos\theta_2'}{1-\beta^2} - 1. \tag{2.13}$$

To exploit the random nature of the process, we average equation 2.13 over angles to end with

$$\frac{\Delta E}{E} = \frac{1+\beta^2/3}{1-\beta^2} - 1 \approx \frac{4}{3}\beta^2, \tag{2.14}$$

where we used $< \cos\theta_2' >= 0$, giving

$$\frac{dn}{d\cos\theta_2'} = \text{const}, -1 \leq \cos\theta_2' \leq 1, \tag{2.15}$$

and the fact that the angular probability of a particle to enter the cloud with an angle θ_1 is proportional to the relative velocity between particle and cloud

$$\frac{dn}{d\cos\theta_1} = \frac{1}{2}(1-\beta\cos\theta_1), -1 \leq \cos\theta_1 \leq 1, \tag{2.16}$$

so that

$$< \cos\theta_1 >= -\frac{\beta}{3}. \tag{2.17}$$

From Eq. 2.14 we get the net energy gain per collision

$$dE \propto \beta^2 E. \tag{2.18}$$

Because of the β^2 dependence this effect is called second order Fermi acceleration. Depending on the angles for a single encounter this process can result in a loss of energy, but after many encounters there is a net energy gain. This is sometimes incorrectly expressed by saying that there are more approaching encounters ($\cos\theta_1 < 0$) than overtaking encounters ($\cos\theta_1 > 0$), but Eq. 2.14 tells us, that an approaching encounter with the cloud in which the particle goes out the back side of the cloud ($\cos\theta_2' < 0$) can result in a loss of energy. Similarly, an overtaking collision can sometimes result in an energy gain. After n encounters the particle has reached the energy

$$E = E_i \cdot e^{\beta^2 n}, \tag{2.19}$$

as can be seen from integrating Eq. 2.18. Here E_i stands for the initial energy with which the particle enters the cloud. To manifest the time dependence in 2.19 we introduce the average time between two collisions τ_c. The number of collision in the interval of time t is then $n = t/\tau_c$. Substituting this last expression in 2.19, we get

$$E(t) = E_i \cdot e^{\beta^2 t/\tau_c} = E_i \cdot e^{t/t_c}, \tag{2.20}$$

where $t_c = \tau_c/\beta^2$. After an average confinement time t_l a CR will be scattered out of the cloud. With $\alpha = t_c/t_l$ the probability that a particle survives until time t is given by

$$P(t) = e^{-t/t_l} = E_i^{t_c/t_l} \cdot (E_i \cdot e^{-t/t_c})^{-t_c/t_l} = k \cdot E^{-\alpha}, \tag{2.21}$$

where we used Equation 2.20. Therefore the total number of particles with energy greater than E is

$$J(E) \propto E^{-\alpha}. \tag{2.22}$$

$J(E)$ shows a typical power law behavior. The differential energy spectrum is proportional to

$$\frac{dJ}{dE} \propto E^{-(\alpha+1)} =: E^{-\gamma}, \tag{2.23}$$

where γ is the spectral index (the spectral index γ is not universal, but depends on the properties of the clouds). Thus, second order Fermi acceleration provides us with a mechanism to explain the observed power law in the local CR fluxes.

However, during their time in the dense clouds particles of course also suffer energy losses by ionization. Consequently, there exists a minimum energy for second order Fermi acceleration, given by the balance of energy losses and energy gains. The mininun energy E_{th} increases with the charge of the particle, because of the increasing ionization losses. For protons this energy is E_{th} = 200 MeV, for oxygen E_{th} = 20 GeV and for iron E_{th} = 300 GeV.

First Order Fermi Acceleration

Another acceleration mechanism was proposed by Bell (1978) and independently by Blandford & Ostriker (1978). Here, a supernova remnant shock front is considered, which generates a non random boost. The basic framework is similar to second order Fermi acceleration, just in this case the cloud is represented by accelerated gas following a shock front (downstream). The shock front moves with velocity $-\mathbf{u_1}$. The shocked gas flows away from the shock with a velocity $\mathbf{u_2}$ relative to the shock front, and $|u_2| < |u_1|$. Thus in the laboratory frame the gas behind the shock moves with $\mathbf{V} = -\mathbf{u_1} + \mathbf{u_2}$. Equation 2.13 also applies to this situation with $\beta = V/c$ now interpreted as the velocity of the shocked gas ("downstream") relative to the unshocked gas ("upstream").

The crucial difference between the two cases comes when we take the angular averages to obtain the average energy gain per encounter. This time the angular distribution 2.15 is replaced by the projection of an isotropic distribution on the shock plane, namely

$$\frac{dn}{d\cos\theta_2'} = 2\cos\theta_2', -1 \leq \cos\theta_2' \leq 0, \tag{2.24}$$

that gives

$$< \cos\theta_2' >= \frac{2}{3}. \tag{2.25}$$

The distribution of $\cos\theta_1$ in 2.16 is again the projection of an isotropic flux onto a plane, but this time with $-1 \leq \cos\theta_1 \leq 0$, so that

$$< \cos\theta_1 >= -\frac{2}{3}. \tag{2.26}$$

Averaging Equation 2.13, we get

$$\frac{\Delta E}{E} = \frac{1 + \frac{4}{3}\beta + \frac{4}{9}\beta^2}{1-\beta^2} - 1 \approx \frac{4}{3}\beta, \tag{2.27}$$

where we assumed that $\beta << 1$, i.e. non relativistic motion of the shock front. The first order dependence on the velocity of the cloud gives the name to this kind of Fermi acceleration. Obviously this type of acceleration mechanism is more efficient than second order Fermi acceleration, because β is always smaller than one. Moreover, for a single encounter the energy gain in the case of first order Fermi acceleration is always positive, because for the infinite plane shock $cos\theta_2'$ is always positive and $cos\theta_1$ is always negative.

The first order Fermi acceleration process predominantly takes place in SN shock fronts (although in the presence of non random Galactic winds in the halo a first order Fermi acceleration is also expected), thus this mechanism defines the source spectrum of CRs. In general the injection spectrum can be parameterized as follows

$$q^j(E) = q_0^j Q^j(E), \tag{2.28}$$

where the j index specifies the nucleus, q_0^j take into account the composition and $Q_j(E)$ encloses the energy dependence. For the last term different forms have been considered. For the HEAO dataEngelmann (1985) it was realized that the best parameterization in the leaky box framework with the proper modulation strength, is

$$\frac{Q^j(E)}{dE} \propto p^{-\gamma}. \tag{2.29}$$

This is in agreement with the law directly derived from the shock acceleration theory (Blandford & Ostriker, 1978). In the case of first order Fermi acceleration the spectral index γ is universal, i.e. it only depends on the Mach number M of the flow. For a mono-atomic gas one gets $\gamma = 2 + 4/M^2$, which for a strong shock with $M >> 1$ is very close to the required value of $\gamma \sim 2.1$ for protons.

The above derivation implicitly assumes that the CR plasma does not affect the acceleration regions. This is called the test particle approach. In reality the CRs being accelerated can cause streaming instabilities and generate hydromagnetic waves, which themselves can the source of diffusion in the upstream, unshocked region. With this coupling the acceleration process is nonlinear and spectra with other indices than the ideal $\gamma \approx 2$ can occur. In particular it is possible to get harder spectra with the energy of the accelerated particles concentrated near the maximum energy of the accelerator.

Maximum energy

The finite lifetime of the supernova blast wave as a strong shock limits the maximum energy per particle that can be achieved with this mechanism. A simple estimate of the maximum possible acceleration rate for nuclei (Drury, 1983; Lagage & Cesarsky, 1983a,b) leads to a maximum energy of

$$E_{max} \leq Z \cdot 3 \cdot 10^4 \text{ GeV}, \tag{2.30}$$

where we assumed $10 M_\odot$ ejected at $5 \cdot 10^8$ cm/s into the nominal ISM with one proton per cubic centimeter and a magnetic field of $B_{ISM} \sim 3\ \mu$G. There are large uncertainties and oversimplifications in the parameters and derivation used to reach Eq. 2.30. By assuming that the SN exploded into the low density, hot ISM, Lagage & Cesarsky (1983b) obtained a somewhat higher maximum energy than in 2.30, but given the uncertainties $E_{max} \sim 100$ TeV is a good round number to use for CR acceleration by SN blast waves. For electrons it is necessary to check whether synchrotron losses give a more restrictive limit for the maximum energy. Comparing the acceleration rate from the derivation of 2.30 with the synchrotron loss rate A.9 gives

$$E_{max}^{synch} \sim 23 \text{ TeV} \frac{u_1}{c} \frac{1}{\sqrt{B}}. \tag{2.31}$$

For the same SN parameters used to obtain 2.30 ($B \sim 3\ \mu$G and $u_1 \sim 5 \cdot 10^8$ cm/s) this gives an upper limit of $\sim$ 220 TeV, nearly a factor ten higher than the limit from SN age in 2.30. Therefore, because of the low magnetic field used in this example, the acceleration is not limited by synchrotron losses. On the other hand it is clear that whenever shock acceleration occurs in a region of high magnetic field, synchrotron losses are likely to limit the acceleration of electrons. Examples for this are neutron stars interacting with nearby matter.

Trap reacceleration in the presence of moving traps

Second order Fermi acceleration as discussed above considers the case of randomly moving magnetic mirrors, such as H_2 clouds which are accompanied by large magnetic fields due to flux freezing. H_2 clouds usually occur in the form of molecular cloud complexes, in which the weak magnetic field in the cloud complexes is focussed into the dense molecular clouds. Two such clouds then constitute a magnetic bottle with the clouds themselves as the focus points. If the distance of the clouds is comparable to or smaller than the mean diffuse scattering length, CRs can be efficiently trapped for a long time (Chandran, 2000, 2001). Zirakashvili (1999) showed that CRs can gain energy due to trap deformation. With δB the value of the random magnetic field, trap reacceleration is a factor $(\delta B/B)^{1/2}$ more efficient than the resonant reacceleration of untrapped particles and might significantly contribute to the total CR reacceleration especially at larger energies (Zirakashvili, 1999).

2.3.3 A Toy Diffusion Model

As discussed in Section 2.1.3 diffusion is the most important effect that characterizes the propagation of CRs. Here we use the diffusion equation in the form of Fick's law to discuss the diffusion of CRs generated by the magnetic turbulences. Note, that one could also use the more general Fokker-Planck equation. Both, Fick's law and the Fokker-Planck equation, are purely phenomenological equations in the sense that they constitute different choices of the flux Γ in the fundamental continuity equation

$$\frac{\partial N}{\partial t} = -\frac{\partial \Gamma}{\partial x}, \tag{2.32}$$

where N is a probability density or number density and $\Gamma = -D(x)\partial_x N$ for Fick's law and $\Gamma = -\partial_x(U(x)N) + \partial_x^2(D(x)N)$ for the Fokker-Planck equation. For a homogeneous system the coefficients U and D are independent of coordinates and different choices for Γ collapse to the same identical form when D and U are constant. For inhomogeneous systems there is no general form of Γ, but Γ has to be chosen according to the microscopic dynamics of the system (e.g. Bian & Garcia (2005) and van Milligen et al. (2005) demonstrate that the straightforward generalization of Fick's law $\Gamma = -D(x)\partial_x N(x)$ cannot hold in all systems). For the cases under consideration (smooth variations on scales much larger than the mean scattering length) the Fokker-Planck equation reduces to Fick's law as discussed in Sattin (2008), to which the reader is referred for further discussion.

CR diffusion is significantly complicated by the fact that the magnetic turbulences CRs scatter on are partially generated by the CRs themselves. Here and in the following we use the test particle approach meaning that the CRs are assumed to move in a static spectrum of turbulences.

A spatial gradient in the density of particles $N(\mathbf{r}, t)$ will generate a current that transports particles from regions of high density to regions of low density

$$\nabla N \neq 0 \rightarrow \mathbf{J}(\mathbf{r}, t) = -D\nabla N, \qquad (2.33)$$

where D is the diffusion coefficient. Note, that no flux is generated for the case of a constant particle distribution, even if the diffusion coefficient depends on spatial coordinates. Naively one might think that regions with decreased diffusion coefficient (i.e. higher density of scattering centers) would lead to an increased CR density, because they confine CRs longer. However, at the same time, regions with increased density of scattering centers act as reflecting surfaces. This can be understood in the following way: for an inhomogeneous density of scattering centers n_{sc}, say, $dn_{sc}/dx < 0$, a test particle at x has a larger probability of striking against a scattering center that is on its left $(x - \delta x)$ rather than on its right $(x + \delta x)$, and therefore of being backscattered in the opposite direction. Hence, there is a larger probability of bouncing back rightwards than the converse. In so far regions with decreased diffusion coefficient will confine CRs for a longer time, but simultaneously CRs are efficiently kept from entering these regions. For systems in which the generalization of Fick's law $\Gamma = -D(x)\partial_x N$ holds, there is no resulting particle flux due to gradients in the diffusion coefficient.

Remind, that *diffusion* occurs even in the case of a homogeneous particle distribution $N(\mathbf{r}, t) = const.$ and a homogeneous distribution of scattering centers. In this case however, the generated *drifts due to diffusion cancel.*

In general we can write the continuity equation with a source term $Q(\mathbf{r}, t)$

$$\frac{\partial N}{\partial t} = -\nabla \cdot \mathbf{J} + Q(\mathbf{r}, t), \tag{2.34}$$

and we immediately get the diffusion equation

$$\frac{\partial N}{\partial t} = \nabla \cdot (D \nabla N) + Q, \tag{2.35}$$

where we used 2.33 to identify the diffusion coefficient. The Green's function associated with 2.35

$$G(\mathbf{r}, t) = \frac{1}{8(\pi D t)^{3/2}} e^{-r^2/(4Dt)}, \tag{2.36}$$

gives us the probability for finding a particle that is injected at $r = 0$ at a position $\mathbf{r}$ after the time t. For a spatially constant diffusion coefficient the mean distance from the Galactic plane at $z = 0$ can be calculated from 2.36

$$< |z| >= \frac{1}{8(\pi D t)^{3/2}} \int z e^{-r^2/(4Dt)} dV = 2\sqrt{Dt/\pi}. \tag{2.37}$$

The characteristic time to reach a height H can then be defined as

$$t_H \equiv \frac{\pi}{4D} H^2 \sim \frac{H^2}{D}. \tag{2.38}$$

With t_H we are able to define the characteristic averaged velocity with which CRs escape from the galaxy with half-halo height H

$$v_D \sim H/t_H \sim D/H. \tag{2.39}$$

It should be underlined that to obtain the mean distance from the galactic plane, we assumed the diffusion coefficient to be spatially constant all over the halo and the disk. Given the different gas densities and the different magnetic field strengths in these two regions this is not necessarily true.

Building a Toy Transport Equation Starting with the phenomenological assumption of diffusion as the main transport process we can build the CR transport equation, by simply choosing the physics we want to include and adding up the relevant terms:

- Diffusion, as derived in the previous paragraph

$$\frac{\partial N_i}{\partial t} = \nabla \cdot (D_i \nabla N_i), \tag{2.40}$$

where $N_i d\epsilon = N_i(t, \mathbf{r}, \epsilon) d\epsilon$ is the number density of i-particles at time t and position $\mathbf{r}$ and ϵ the energy per nucleon;

- In the presence of systematic large-scale motion of the medium such as Galactic convection, we have to consider the term

$$\frac{\partial N_i}{\partial t} = -N_i \vec{\nabla} \cdot \mathbf{u}, \tag{2.41}$$

but for now we assume this effect to be absent;

- Continuous energy losses can be introduced collectively by defining the energy loss per unit time $b_i = d\epsilon/dt$ for each particle

$$\frac{\partial N_i}{\partial t} = -\frac{\partial}{\partial \epsilon_k}(b_i N_i); \tag{2.42}$$

- Inelastic scattering with the interstellar medium

$$\frac{\partial N_i}{\partial t} = -nv\sigma_i N_i, \tag{2.43}$$

where n is the gas density of the ISM, v is the velocity of the particle and σ_i is the inelastic scattering cross section of a nucleus of type i with the interstellar gas;

- Production from inelastic scattering of heavier nuclei can be included by

$$\frac{\partial N_i}{\partial t} = \sum_{i<j} nv\sigma_{ij} N_j, \tag{2.44}$$

where σ_{ij} is the production cross section of nucleus i from the heavier nucleus j;

- Radioactive decay is described by

$$\frac{\partial N_i}{\partial t} = -\frac{N_i}{\tau_i}, \tag{2.45}$$

where τ_i is the lifetime of the i-nucleus;

- Production from heavier nuclei by radioactive decays is given by

$$\frac{\partial N_i}{\partial t} = \sum_{i<j} \frac{N_j}{\tau_{ij}}, \tag{2.46}$$

with τ_{ij} representing the probability that a nucleus of type j decays into a nucleus of type i.

Adding up the above processes we end up with a very simple but comprehensible propagation equation describing a diffusion model

$$\frac{\partial N_i}{\partial t} = q_i + \nabla \cdot (D_i \nabla N_i) - \frac{\partial}{\partial t}(b_i N_i) - (nv\sigma_i + \frac{1}{\tau_i})N_i + \sum_{i<j}(nv\sigma_{ij} + \frac{1}{\tau_{ij}})N_j, \tag{2.47}$$

where $q_i = q_i(t, \mathbf{r}, \epsilon_k)$ is the source function that describes the power and space-time distribution of point like sources producing the nuclei of type i.
Equation 2.47 can be solved analytically for a variety of simplificating assumptions. In order to decouple the diffusion of particles from the fragmentation we introduce an integral form for the solution

$$N_i(t, \mathbf{r}) = \int_0^\infty N_i(x) G(t, \mathbf{r}, x) dx, \tag{2.48}$$

where $G(t, \mathbf{r}, x)$ is the path-length distribution function describing the fraction of particles at time t and point $\mathbf{r}$ which have traversed a layer of matter of thickness x. In addition we assume that

- ionization losses are absent;
- the sources are the same for all kind of nuclei but the production of each nucleus is weighted by the constant coefficient q_i

$$q_i(\mathbf{r}, t) = g_i \chi(t, \mathbf{r}) \tag{2.49}$$

- the diffusion tensor does not depend on the kind of nuclei.

By plugging 2.48 into Equation 2.47 we get the system of equations

$$\frac{\partial G}{\partial x} + \frac{\partial G}{\partial t} - \nabla \cdot (D \nabla G) = \chi \delta(x),$$
$$\frac{\partial N_i}{\partial x} = -(nv\sigma_i + \frac{1}{\tau_i})N_i + \sum_{i<j}(nv\sigma_{ij} + \frac{1}{\tau_{ij}})N_k + g_i \delta(x), \tag{2.50}$$

provided that the following initial conditions are satisfied

$$[G(t,\mathbf{r},0)+\chi(t,\mathbf{r},0)][N_i(0)+g_i]=0,\ \ G(t,\mathbf{r},\infty)N_i(\infty)=0. \tag{2.51}$$

The first equation in 2.50 describes diffusion while the second equation describes fragmentation and decay and is sometimes considered alone in the so-called slab model. This last equation is a first order differential equation and we can write the solution as a sum of exponentials

$$N_i(x)=\sum_{j=1}^{i} a_{ij}e^{-(nv\sigma_i+\frac{1}{\tau})x}, \tag{2.52}$$

where the coefficient a_{ij} can be determined by solving recurrence relations obtained by substituting 2.52 into the starting equation 2.50. Substituting back 2.52 in 2.48 we get the final solution which is

$$N_i(t,\mathbf{r})=\sum_{j=1}^{i} a_{ij}F_j(t,\mathbf{r}), \tag{2.53}$$

where we introduced the Laplace transformation in the variable $nv\sigma_i+\frac{1}{\tau}$

$$F_i(t,\mathbf{r})=\int_0^\infty G(t,\mathbf{r},x)e^{-(nv\sigma_i+\frac{1}{\tau_i})x}dx. \tag{2.54}$$

Thus the particle density is determined by $F_i(t,\mathbf{r})$, which is a solution of the equation obtained by multiplying each term in 2.50 by $\exp-(nv\sigma_i+1/\tau_i)$ and integrating over dx, namely

$$\frac{\partial F_i}{\partial t}-\nabla\cdot(D\nabla F_i)+(nv\sigma_i+\frac{1}{\tau_i})F_i=0, \tag{2.55}$$

where we used

$$\int_0^\infty \frac{\partial G}{\partial x}e^{-(nv\sigma_i+\frac{1}{\tau_i})x}=-G(t,\mathbf{r},0)+nv\sigma_i+\frac{1}{\tau_i}F_i,\ \ G(t,\mathbf{r},0)=-\chi(t,\mathbf{r},0). \tag{2.56}$$

$F(t,\mathbf{r})$ is a momentum generating function for the mean path-length traversed by the particles

$$<x>=\frac{\int_0^\infty xGdx}{\int_0^\infty Gdx}=-\left(\frac{d}{d(nv\sigma_i+1/\tau_i)}\ln F_i\right)|_{\sigma_i=0}, \tag{2.57}$$

Equation 2.47 offers the possibility to find analytical solutions under a number of simplifying assumptions. This equation forms the basis for a variety of contemporary models such as the DarkSUSY (Gondolo et al., 2004) code or the model employed by Donato et al. (2002). However, remind that we build this equation by just adding up terms that we considered relevant for CR transport. Equation 2.47 is by no means a complete description of the physics of CR transport since it does not include any reacceleration processes and it neglects convection. In addition, most handy (and therefore widely-used) solutions have to make additional simplifying assumptions about the geometry of the Galaxy, which makes the model predictions very sketchy.

2.3.4 Leaky Box Models

An extremely simplified, yet helpful, approximation of a diffusion model is the so called "leaky-box" approximation. Although the predictive power of this model is rather limited, it provides us with some very useful tools that can help to understand CR physics on a very illustrative level and will come in handy in more complicated cases later on. Here we will discuss the leaky-box model focussing on its predictions for radioactive instable isotopes and secondary to primary ratios. The predictions of this simple model form the motivation for us to derive the full CR transport equation in the diffusion limits in Section 2.4.

Leaky-box models can be understood as a limit of a diffusion model assuming that

1. scattering in the disk and halo is weak and there is strong reflection only at the galactic boundary and
2. there is only little leakage from the Galaxy.

The first assumption forces particles to traverse the Galactic plane many times before escaping to intergalactic space, as required by the observed grammage (see Section 2.1.3). The second assumption allows us to consider only spatially averaged quantities, thus leading to spatially constant CR densities and escape times. Physically, this approximation is justified by the argument that diffusion is fast. In this case we can express the diffusion term as

$$\nabla(D_i \nabla N_i) \longrightarrow -\frac{N_i}{\tau_{esc}}, \tag{2.58}$$

where τ_{esc} is the confinement time for CRs, given by the diffusion coefficient $D \sim \kappa(p)$:

$$\tau_{esc}(p) \propto \kappa^{-1}(p). \tag{2.59}$$

After this substitution one can assume that CRs stream freely inside the confinement volume, with reflections only at the boundaries. Note, that in the case of free-streaming no reacceleration can occur, so in the leaky-box approximation the neglection of these terms is implicitly justified. The transport equation then reads

$$\frac{\partial N_i}{\partial t} = q_i - \frac{N_i}{\tau_{esc}} - \frac{\partial(b_i N_i)}{\partial \epsilon_k} - (n_H v \sigma_i + \frac{1}{\tau_i})N_i + \sum_{i<j}(n_H v \sigma_{ij} + \frac{1}{\tau_{ij}})N_j. \tag{2.60}$$

For CR nuclei with energies above a few MeV/nucleon we can neglect any continuous momentum losses. Asking only for the steady state solution the transport equation then simplifies to

$$(n_H v \sigma_i + \frac{1}{\tau_i})N_i = q_i - \frac{N_i}{\tau_{esc}} + \sum_{j<i}(n_H v \sigma_{ij} + \frac{1}{\tau_{ij}})N_j. \tag{2.61}$$

Despite the many assumptions we made, this equation is very useful for simple estimates, e.g. the escape time of CRs or the halo height. In the following we will apply the leaky-box approximation to the case of secondary to primary ratios and radioactive isotopes.

Secondary to primary ratios Following Schlickeiser (2002) we now apply equation 2.61 to a secondary CR element (e.g. boron) that mainly results from the spallation of primary elements (in the case of B predominantly C, N and O). Any additional contribution from radioactive decays of heavier nuclei such as ^{10}Be is negligible. The secondary source is therefore given by

$$\sum_{j>B}(n_H v \sigma_{Bj} + \frac{1}{\tau_{Bj}})N_j \simeq \sigma_B v n_H N_{CNO}, \tag{2.62}$$

where σ_B denotes the partial fragmentation cross section for the production by CNO-nuclei. Since B is stable and can only be lost by spallation or escape from the Galaxy the transport equation for boron then reads

$$(n_H v \sigma_{f,B} + \frac{1}{\tau_{esc}})N_B = \sigma_B v n_H N_{CNO}, \tag{2.63}$$

where $\sigma_{f,B}$ is the cross section for catastrophic loss of boron due to spallation. Equation 2.63 then yields for the secondary to primary ratio

$$\frac{N_B}{N_{CNO}} = \frac{\sigma_B}{\sigma_{f,B} + (v n_H \tau_{esc})^{-1}}. \tag{2.64}$$

In the leaky-box approximation with the homogeneous confinement volume the grammage reads as

$$X(p) = \int_0^\infty dl\ n(\mathbf{r}) \simeq v n_H \tau_{esc}, \tag{2.65}$$

where v is the CRs velocity, n_H the average gas density in the Galaxy and τ_{esc} the mean residence time of the primary CR in the Galaxy. Now we can write equation 2.64 as

$$\frac{N_B}{N_{CNO}} = \frac{\sigma_B}{\sigma_{f,B} + X^{-1}(p)}. \tag{2.66}$$

The power law decrease of the measured secondary/primary ratios $\propto p^{-0.5}$ at relativistic particle energies shown in Fig. 2.3 indicates according to 2.66 a corresponding decrease of the CR escape time and according to equation 2.59 a corresponding momentum variation of the spatial diffusion coefficient

$$\kappa(p) \propto p^{0.5}. \tag{2.67}$$

Such a momentum variation is possible for Kolmogorov turbulence when the turbulence spectral index $q = 2 - b \simeq 1.5$ is close to the Kraichnan value. However, one has to keep in mind that this simple estimate is only valid in the leaky-box approximation. A more detailed model including the effects of reacceleration will give a different value.

Secondary Cosmic Ray Clocks In order to get an estimate of the time CRs spend in the confinement volume we consider two isotopes of one secondary CR element, one of which is radioactive instable (e.g. ^{10}Be) and the other one is stable (e.g. ^{9}Be). Since beryllium is a secondary CR element, produced by spallation of CNO-nuclei in the interstellar medium, the respective source terms are

$$q_9(p) = \sigma_9 v n_h N_{CNO} \text{ and } q_{10}(p) = \sigma_{10} v n_h N_{CNO}, \tag{2.68}$$

where the partial spallation cross sections for the channels $(CNO) \rightarrow^9 Be$ and $(CNO) \rightarrow^{10} Be$ are known from accelerator experiments.
The catastrophic loss times due to spallation can be expressed as $\tau_{c,9} = (v n_H \sigma_{f,9})^{-1}$and $\tau_{c,10} = (v n_H \sigma_{f,10})^{-1}$ respectively. The transport equation 2.61 for the stable ^9Be isotope then reads

$$N_9(v n_H \sigma_{f,9} + \frac{1}{\tau_{esc}(p)}) = \sigma_9 v n_H N_{CNO}, \tag{2.69}$$

while the transport equation for the unstable ^{10}Be isotope reads

$$N_{10}(vn_H\sigma_{f,10} + \frac{1}{\tau_{esc}(p)} + \frac{1}{\gamma\tau_d}) = \sigma_{10}vn_H N_{CNO}. \quad (2.70)$$

The ratio of the two isotopes then becomes

$$\frac{N_{10}}{N_9} = \frac{\sigma_{10}}{\sigma 9} \frac{vn_H\sigma_{f,9} + \tau_{esc^{-1}}(p)}{vn_H\sigma_{f,10} + \tau_{esc^{-1}}(p) + (\gamma\tau_d)^{-1}}. \quad (2.71)$$

All cross sections and the decay lifetimes in this equation are known and thus by measuring the ratio 2.71 at different energies one can infer the value of the gas density in the confinement region n_H and the value and the momentum dependence of the escape time $\tau_{esc}(p)$. By combining with inferences drawn from the traversed grammage (Eq. 2.66) from secondary to primary measurements one can determine the values of n_H and τ_{esc} separately. Table 2.1 summarizes the density and escape time measured with several cosmic ray isotopes. All measurements are consistent with a CR escape time of $\simeq 10^7$ yrs at non-relativistic particle energies. The average gas density seen by CRs is substantially lower than the Galactic average in the disk of 1 atom cm^{-3}, which indicates that CRs have spent substantial parts of their lifetime in low density regions of the interstellar medium, like, e.g. the halo of the Milky Way.
The time scale for continuous momentum losses of relativistic CR nucleons in the Galaxy is much longer than $10^7 yr$ (see Fig. 2.6). Thus, the neglection of these processes in the above approximation of the transport equation 2.61 is justified.

2.4 The Cosmic Ray Transport Equation

The constraints from secondary CRs and radioactive isotopes found in the last section provide good reasons to model CR transport by diffusion. However, up to now the description was purely phenomenological and the underlying physical processes remained unknown.
In this section we will derive the full CR transport equation, starting with the Boltzmann kinetic equation. To this end we work in the context of the kinetic theory following the derivation of the propagation equation given in Ginzburg et al. (1990) to which we refer the reader for further explanations.
At the end of this section stands the full transport equation for CRs in the diffusion-convection approximation, which will serve as the basis for all CR transport models in the

Isotope	ρ [atoms x cm^3]	CR lifetime [Myr]	Reference
^{10}Be	0.18(+0.18,-0.11)	17(+24,-8)	Garcia-Munoz, Mason, Simpson '77
	0.30(+0.12,-0.10)	8.4(+0.4,-2.4)	Wiedenbeck,Greiner '80
	0.23(+0.13,-0.11)	14(+13,-5)	Garcia-Munoz, Simpson, Wefel '81
	0.24(+0.07,-0.07)	15(+7,-4)	Simpson, Garcia-Munoz '88
	0.28(+0.14,-0.11)	27(+19,-9)	Lukasiak et al. '94
	0.23(+0.04,-0.04)	18(+3,-3)	Connell '97
^{26}Al	0.28(+0.72,-0.19)	9(+20,-6.5)	Wiedenbeck '83
	0.52(+0.26,-0.20)	13.5(+8.5,-4.5)	Lukasiak, McDonald, Webber '94
	0.28(+0.72,-0.19)	15.6(+2.5,-2.6)	Connell, Simpson '97
^{36}Cl	0.39(+0.15,-0.15)	11(+4,-4)	Connell & et al. (1997)
^{54}Mn	0.37(+0.16,-0.11)	14(+6,-4)	DuVernois '97

Table 2.1: Density and escape time measurements of cosmic clocks. Adapted from Schlickeiser (2002). N.B.: (Adapted from DuVernois (1997)). These are the quoted, published, confinement times. Using a different pathlength distribution (PLD) would alter these values somewhat. Mn confinement is for an assumed ^{54}Mn β^+ partial half life of 1 Myr.

remainder of this thesis.

2.4.1 The Boltzmann Kinetic Equation

The number of particles contained in the phase space volume Γ at the time t can be written as the integral of the phase space density distribution $f(t, \mathbf{x}, \mathbf{p})$

$$N(t) = \int\limits_{\Gamma} d\mathbf{x}d\mathbf{p} f(t, \mathbf{x}, \mathbf{p}) = \int\limits_{\Gamma} dn. \tag{2.72}$$

During an infinitesimal timestep δt, the position and momentum of the particles change by

$$\begin{aligned} \delta\mathbf{x} &= \dot{\mathbf{x}}\delta t = \frac{\mathbf{p}}{m}\delta t, \\ \delta\mathbf{p} &= \dot{\mathbf{p}}\delta t = \frac{\mathbf{F}}{m}\delta t, \end{aligned} \tag{2.73}$$

where **F** is an external force. The total number of particles in the infinitesimal phase space volume $\delta\mathbf{x}\delta\mathbf{p}$ then changes by

$$\delta dn = d\mathbf{x}d\mathbf{p}\;\delta f(t,\mathbf{x},\mathbf{p}) = d\mathbf{x}d\mathbf{p}\left(\frac{\partial f}{\partial t} + \dot{\mathbf{x}}\nabla f + \dot{\mathbf{p}}\frac{\partial f}{\partial \mathbf{p}}\right)\delta t. \tag{2.74}$$

In the absence of energy losses through collisions the Liouville theorem holds, asserting that

$$\delta f(t,\mathbf{x},\mathbf{p}) = 0. \tag{2.75}$$

If collisions occur, we have to introduce the variation rate R of the distribution function

$$dn' - dn = R\delta t, \tag{2.76}$$

where the rate R has to be specified case by case, depending on the properties of each interaction. With 2.74 and 2.76 the full kinetic equation reads

$$\frac{\partial f}{\partial t} + \dot{\mathbf{x}}\nabla f + \dot{\mathbf{p}}\frac{\partial f}{\partial \mathbf{p}} = R. \tag{2.77}$$

Collisions between particles are always mediated by an interaction of some sort. Consequently, we have to distinguish between forces whose contribution can be included in the collision term R, and forces which contribute to the $\dot{\mathbf{p}}$ term in 2.77. The two relevant quantities to compare are the interaction range and the mean free path of the particles. In a gas of neutral atoms, e.g., the range of interactions extends about distances of the order of atomic dimensions. Compared to the mean free path these distances are very short. In this case we have a genuine collision framework and if no external field is applied the $\dot{p}$ term in 2.77 can be neglected.

CRs form a plasma of ionized particles and each particle carries a charge which is responsible for a long range interaction. The collective macroscopic field, generated by the CRs themselves, reduces the free mean path to zero. Thus, if the plasma is dense enough we can neglect collisions. In the case of a rarefied plasma the Debye length can be used to characterize the interaction range of the macroscopic field. If the Debye length is large compared to the mean distance between particles, collisions can be disregarded. Introducing the Lorentz force $\mathbf{F} = Ze\left(\vec{\mathcal{E}} + (\dot{\mathbf{x}} \times \vec{\mathcal{H}})/c\right)$, the magnetic field $\vec{\mathcal{H}}$ and the electric field

$\vec{\mathcal{E}}$, we can write the kinetic equation as

$$\frac{\partial f}{\partial t} + \dot{\mathbf{x}}\nabla f + Ze\left(\vec{\mathcal{E}} + \frac{\dot{\mathbf{x}} \times \vec{\mathcal{H}}}{c}\right)\frac{\partial f}{\partial \mathbf{p}} = 0. \tag{2.78}$$

2.4.2 Quasi-Linear Approximation

The assumption that the distribution $f(t, \mathbf{x}, \mathbf{p})$ is dominated by a slowly evolving part and only a small fluctuating part is called the standard quasi-linear approximation. In this case $f(t, \mathbf{x}, \mathbf{p})$ can be expressed as

$$f(t, \mathbf{x}, \mathbf{p}) = f_0(\mathbf{p}) + f_1(t, \mathbf{x}, \mathbf{p}), \tag{2.79}$$

where $f_0(\mathbf{p})$ is the slowly evolving background and $f_1(t, \mathbf{x}, \mathbf{p})$ the fluctuating part which is required to be small with respect to f_0. We consider an instability such that a continuous spectrum of waves is excited. If the wavelength of the perturbation is of the order of a few Debye lengths f_0 can be considered spatially uniform. Averaging 2.79 over the spectrum of perturbations the fluctuating part vanishes

$$< f_1 >= 0, \tag{2.80}$$

and we are left with

$$< f >= f_0. \tag{2.81}$$

The same procedure is applied to the magnetic field, so that

$$\vec{\mathcal{H}} = \vec{\mathcal{H}}_0 + \vec{\mathcal{H}}_1, \tag{2.82}$$

where $\vec{\mathcal{H}}_1$ represents the random fluctuating part of $\vec{\mathcal{H}}$ and

$$< \vec{\mathcal{H}} >= \vec{\mathcal{H}}_0, \; < \vec{\mathcal{H}}_1 >= 0. \tag{2.83}$$

Since the CR plasma is a highly conductive medium, the mean value of the electric field $\vec{\mathcal{E}}$ over the ensemble of waves vanishes, so that

$$< \vec{\mathcal{E}} >= 0. \tag{2.84}$$

The perturbations $\vec{\mathcal{H}}_1$ and $\vec{\mathcal{E}}$ can be thought of as a superposition of waves with random continuous phases. Using a Fourier expansion we get

$$\vec{\mathcal{H}}_1 = \sum_\alpha \int d^3k e^{-i[\omega^\alpha(\mathbf{k})t - \mathbf{k}\cdot\mathbf{r}]} \vec{\mathcal{H}}_1^\alpha(\mathbf{k}),$$
$$\vec{\mathcal{E}} = \sum_\alpha \int d^3k e^{-i[\omega^\alpha(\mathbf{k})t - \mathbf{k}\cdot\mathbf{r}]} \vec{\mathcal{E}}^\alpha(\mathbf{k}), \tag{2.85}$$

where the index α refers to different types of waves. Averaging the homogeneous Maxwell equation

$$\nabla \times \vec{\mathcal{E}} = -\frac{1}{c}\frac{\partial \vec{\mathcal{H}}}{\partial t}, \tag{2.86}$$

and applying Eqs. 2.82 and 2.84 we have

$$\nabla \times \vec{\mathcal{E}} = -\frac{1}{c}\frac{\partial \vec{\mathcal{H}}_1}{\partial t}. \tag{2.87}$$

This equation allows us to express the electric and magnetic Fourier coefficients 2.85 in terms of one another by

$$\vec{\mathcal{H}}_1^\alpha(\mathbf{k}) = \frac{c}{\omega^\alpha(\mathbf{k})}[\mathbf{k} \times \vec{\mathcal{E}}^\alpha(\mathbf{k})], \tag{2.88}$$

thus enabling us to eliminate one type of coefficient.

2.4.3 Approximated Solution for the Slowly Varying Distribution

We now go back to equation 2.78 and use a slightly relaxed but more realistic quasi-linear approximation to solve it. In this case the dependence of f_0 on space and time is not neglected, but the fluctuating parts f_1, $\vec{\mathcal{E}}$ and $\vec{\mathcal{H}}_1$ are still assumed to be small compared to equilibrium values. The goal is to find a closed differential equation for f_0 taking into account the influence of the perturbations $\vec{\mathcal{E}}$ and $\vec{\mathcal{H}}_0$. The searched equation will form the basis for our subsequent diffusion model. In a first step we look for a first order expression for f_1 satisfying a differential equation approximated up to linear terms in f_1, $\vec{\mathcal{E}}$ and $\vec{\mathcal{H}}_1$. In a second step this first order expression can be used to close a second order averaged differential equation for f_0.

Inserting equations 2.79 and 2.82 into the Boltzmann equation 2.78 we get

$$\frac{\partial(f_0+f_1)}{\partial t}+\left(\dot{\mathbf{x}}\cdot\vec{\nabla}\right)(f_0+f_1)+Ze\left[\vec{\mathcal{E}}+\frac{\dot{\mathbf{x}}}{c}\times\left(\vec{\mathcal{H}}_0+\vec{\mathcal{H}}_1\right)\right]\frac{\partial f_0}{\partial\mathbf{p}}+$$
$$+Ze\left[\vec{\mathcal{E}}+\frac{\dot{\mathbf{x}}}{c}\times\left(\vec{\mathcal{H}}_0+\vec{\mathcal{H}}_1\right)\right]\frac{\partial f_1}{\partial\mathbf{p}}=0. \tag{2.89}$$

Averaging equation 2.89 yields

$$\frac{\partial f_0}{\partial t}+\left(\dot{\mathbf{x}}\cdot\vec{\nabla}\right)f_o+Ze\left(\frac{\dot{\mathbf{x}}}{c}\times\vec{\mathcal{H}}_0\right)\frac{\partial f_0}{\partial\mathbf{p}}=-<Ze\left(\vec{\mathcal{E}}+\frac{\dot{\mathbf{x}}}{c}\times\vec{\mathcal{H}}_1\right)\frac{\partial f_1}{\partial\mathbf{p}}>, \tag{2.90}$$

where 2.80, 2.83 and 2.84 have been used. Neglecting all the quadratic terms in f_1, $\vec{\mathcal{E}}$ and $\vec{\mathcal{H}}_1$, equations 2.89 and 2.90 reduce to

$$\frac{\partial(f_0+f_1)}{\partial t}+\left(\dot{\mathbf{x}}\cdot\vec{\nabla}\right)(f_0+f_1)+Ze\left[\vec{\mathcal{E}}+\frac{\dot{\mathbf{x}}}{c}\times\left(\vec{\mathcal{H}}_0+\vec{\mathcal{H}}_1\right)\right]\frac{\partial f_0}{\partial\mathbf{p}}+$$
$$+Ze\left[\frac{\dot{\mathbf{x}}}{c}\times\vec{\mathcal{H}}_0\right]\frac{\partial f_1}{\partial\mathbf{p}}=0, \tag{2.91}$$
$$\frac{\partial f_0}{\partial t}+\left(\dot{\mathbf{x}}\cdot\vec{\nabla}\right)f_o+Ze\left(\frac{\dot{\mathbf{x}}}{c}\times\vec{\mathcal{H}}_0\right)\frac{\partial f_0}{\partial\mathbf{p}}=0. \tag{2.92}$$

Now, subtracting Eq. 2.92 from Eq. 2.91, we get a closed expression for f_1

$$\frac{\partial f_1}{\partial t}+\left(\dot{\mathbf{x}}\cdot\vec{\nabla}\right)f_1+Ze\left(\frac{\dot{\mathbf{x}}}{c}\times\vec{\mathcal{H}}_0\right)\frac{\partial f_1}{\partial\mathbf{p}}=-Ze\left(\vec{\mathcal{E}}+\frac{\dot{\mathbf{x}}}{c}\times\vec{\mathcal{H}}_1\right)\frac{\partial f_0}{\partial\mathbf{p}}, \tag{2.93}$$

This step is equivalent to assuming that f_0 depends only on $\mathbf{p}$, since the same result is achieved directly by inserting $f_0 = f_0(\mathbf{p})$ in Eq. 2.91. As a solution of the first order approximation 2.93 the expression

$$f_1=-\int_{-\infty}^{t}dt'Ze\left(\vec{\mathcal{E}}+\frac{\dot{\mathbf{x}}}{c}\times\vec{\mathcal{H}}_1\right)\frac{\partial f_0}{\partial\mathbf{p}}, \tag{2.94}$$

can be used. This is the first order expression for f_1 we searched for. Inserting this into the the second order differential equation 2.90 yields the closed differential equation for f_0

we were looking for

$$
\begin{aligned}
&\frac{\partial f_0}{\partial t} + \left(\dot{\mathbf{x}} \cdot \vec{\nabla}\right) f_0 + Ze\left(\frac{\dot{\mathbf{x}}}{c} \times \vec{\mathcal{H}}_0\right) \frac{\partial f_0}{\partial \mathbf{p}} = \\
&= < Ze\left(\vec{\mathcal{E}} + \frac{\dot{\mathbf{x}}}{c} \times \vec{\mathcal{H}}_1\right) \frac{\partial}{\partial \mathbf{p}} \int_{-\infty}^{t} dt' Ze\left(\vec{\mathcal{E}} + \frac{\dot{\mathbf{x}}}{c} \times \vec{\mathcal{H}}_1\right) \frac{\partial f_0}{\partial \mathbf{p}} > . \qquad (2.95)
\end{aligned}
$$

To simplify this expression we introduce cylindrical coordinates $(p_{||}, p_{\perp}, \varphi)$ in momentum space with

$$
\mathbf{p} \cdot \vec{\mathcal{H}}_0 = p_{||} H_0. \qquad (2.96)
$$

The time scale Δt of particle motion in a magnetic field is given by the gyro-frequency ω_H associated to the circular motion of a particle of charge Ze induced by the surrounding magnetic field, namely

$$
\Delta t^{-1} \sim \omega_H = \frac{ZeHc}{E}, \qquad (2.97)
$$

where E is the particle's total energy

$$
E^2 = \mathbf{p}^2 c^2 + m^2 c^4. \qquad (2.98)
$$

In the case of large turbulence the time scale Δt is much larger than the time scale of fluctuations in $\vec{\mathcal{E}}$ and $\vec{\mathcal{H}}_1$, so that CR transport is driven by the random field fluctuations. In the limit of weak turbulence, corresponding to the assumption that the time scale Δt is much smaller than the time scale associated to the frequency of the random fluctuations of $\vec{\mathcal{E}}$ and $\vec{\mathcal{H}}_1$, the field fluctuations have no effect on the circular motion of fast moving particles. This allows a reasonable average over the angle φ so that $f_0(t, \mathbf{r}, \mathbf{p})$ can be replaced by

$$
\bar{f}_0(t, \mathbf{r}, p_{||}, p_{\perp}) = \frac{1}{2} \int_0^{2\pi} d\varphi < f_0(t, \mathbf{r}, p_{||}, p_{\perp}) > . \qquad (2.99)
$$

It can be demonstrated (for more details see Akhiezer (1975) and Kennel & Engelmann (1966)) that with the above approximation, equation 2.95 can be written as

$$
\begin{aligned}
&\frac{\partial \bar{f}_0}{\partial t} + v_{||} \frac{\partial \bar{f}_0}{\partial z} = \\
= \; &\pi Z^2 e^2 \sum_{\alpha} \int d^3k \sum_{s=-\infty}^{\infty} \left\langle \left[\mathcal{E}_{||}^{\alpha} J_s \hat{\mathcal{P}}_{||}^{\alpha} + \mathcal{E}_{\perp}^{\alpha} \hat{\mathcal{P}}_{\perp}^{\alpha} + \frac{\mathcal{E}_{\perp}^{\alpha}}{p_{\perp}} \left(1 - \frac{k_{||} v_{||}}{\omega^{\alpha}(\mathbf{k})}\right) - \frac{\mathcal{E}_{||}^{\alpha}}{p_{\perp}} \frac{v_{||}}{v_{\perp}} \frac{s\omega_H}{\omega^{\alpha}(\mathbf{k})} J_s\right]^* \cdot \right. \\
\cdot \; &\left. \left[\mathcal{E}_{||}^{\alpha} J_s \hat{\mathcal{P}}_{||}^{\alpha} + \mathcal{E}_{||}^{\alpha} \hat{\mathcal{P}}_{\perp}^{\alpha}\right] \cdot \delta(\omega^{\alpha}(\mathbf{k}) - k_{||} v_{||} - s\omega_H) \right\rangle \bar{f}_0, \qquad (2.100)
\end{aligned}
$$

where

$$\hat{\mathcal{P}}_{||}^{\alpha} = \frac{\partial}{\partial p_{||}} - \frac{s\omega_H}{\omega_\alpha(\mathbf{k})}\frac{1}{v_\perp}\left(v_\perp \frac{\partial}{\partial p_{||}} - v_{||}\frac{\partial}{\partial p_\perp}\right), \tag{2.101}$$

$$\hat{\mathcal{P}}_{\perp}^{\alpha} = \frac{\partial}{\partial p_\perp} - \frac{\mathbf{k}_{||}}{\omega_\alpha(\mathbf{k})}\left(v_\perp \frac{\partial}{\partial p_{||}} - v_{||}\frac{\partial}{\partial p_\perp}\right), \tag{2.102}$$

$$\mathcal{E}_{\perp}^{\alpha} = \frac{1}{2}\left(\mathcal{E}_R^{\alpha}(\mathbf{k})e^{i\Psi}J_{s+1} + \mathcal{E}_L^{\alpha}(\mathbf{k})e^{-i\Psi}J_{s-1}\right), \tag{2.103}$$

$$\mathcal{E}_{L,R}^{\alpha}(\mathbf{k}) = \mathcal{E}_x^{\alpha}(\mathbf{k}) \pm \mathcal{E}_y^{\alpha}(\mathbf{k}), \tag{2.104}$$

$$\mathcal{E}_{||}^{\alpha}(\mathbf{k}) = \mathcal{E}_z^{\alpha}(\mathbf{k}), \tag{2.105}$$

and $\mathcal{E}_{||}^{\alpha}$ is the component of $\vec{\mathcal{E}}^{\alpha}$ projected along H_0, Ψ is the azimuthal angle of the wave vector and $J_s(k_\perp v_\perp/\omega_H)$ the Bessel function of order s. The resonance character of the particle-wave interaction is introduced by the δ-function in 2.100. Resonance is realized by the condition

$$\omega^{\alpha}(\mathbf{k}) = k_{||}v_{||} + s\omega_H,\ s \in \mathbb{Z}, \tag{2.106}$$

where $k_{||}v_{||}$ takes into account the Doppler effect and $s\omega_H$ stands for the cyclotron rotation in the magnetic field $\mathcal{H}_{\prime}$.

A comparison between the particle's Lamor radius $r_H = v_\perp/|\omega_H|$, which is the curvature induced by the magnetic field H_0, and the wavelength of turbulence tells us whether the particle can interact with the wave. In this case the particle is said to be magnetized. There are two limits to discuss

- magnetized particle: $k_\perp v_\perp/\omega_H << 1 \Rightarrow$ the harmonics involved are
  1. $s = 0$ for $\mathcal{E}_{||}^{\alpha} \neq 0 \Rightarrow \omega^{\alpha} = k_{||}v_{||}$;
  2. $s = \pm 1$ for $\mathcal{E}_{\perp}^{\alpha} \neq 0 \Rightarrow w^{\alpha}(\mathbf{k}) = k_{||}v_{||} \pm \omega_H$;
- unmagnetized particle: $k_\perp v_\perp/\omega_H >> 1 \Rightarrow$ the harmonics involved are $s \sim k_\perp v_\perp/\omega_H \Rightarrow \omega^{\alpha}(\mathbf{k}) - k_{||}v_{||} - s\omega_H \sim \omega^{\alpha}(\mathbf{k}) - \mathbf{k}\cdot\mathbf{v} = 0$.

In the second case we have short magnetized waves with $\lambda << 2\pi r_H$, where the effective scattering rate of ultra-relativistic particles has an energy dependence E^{-2}. Such an energy dependence is ruled out by the slope of the B/C ratio above 1 GeV, which decreases as $\approx E^{-\mu}$ with $\mu = 0.5$ (see Fig. 2.3). Such a decrease implies a corresponding decrease in the escape time

$$T \propto E^{-\mu},\ \mu = 0.5. \tag{2.107}$$

If the wavelength is comparable to the Lamor radius, i.e. $\lambda \sim 2\pi r_H$, each frequency in the spectrum of turbulences interacts with a particle of different energy, leading to an energy dependence of the effective scattering rate and thus the diffusion coefficient. A full discussion would require to consider different types of waves separately. Here we restrict ourselves to the case

$$v_s << v_A, \tag{2.108}$$

with

$$v_s = \sqrt{\frac{KT_e}{m}}, \tag{2.109}$$

the sound speed in a medium with T_e being the temperature of thermal electrons and

$$v_A = \frac{H_0}{\sqrt{4\pi\rho}}, \tag{2.110}$$

the Alfvèn velocity. ρ is the density of the medium, i.e. the CR density. With this assumption there are only two types of oscillations left in the magnetohydrodynamic region ($\omega << \omega_H$):

- Alfvèn waves identified by the dispersion relation

$$\omega^\alpha(\mathbf{k}) = \pm|k_{||}|v_A, \tag{2.111}$$

  and by the following properties

$$\begin{aligned} &\mathbf{v} \perp (\mathbf{k}, \vec{\mathcal{H}}_0), \\ &\vec{\mathcal{E}} \in (\mathbf{k}, \vec{\mathcal{H}}_0), \end{aligned} \tag{2.112}$$

  where $(\mathbf{k}, \vec{\mathcal{H}}_0)$ is the plane identified by $\mathbf{k}$ and $\vec{\mathcal{H}}_0$;

- magnetosonic waves identified by the dispersion relation

$$\omega^\alpha(\mathbf{k}) = \pm k v_A, \tag{2.113}$$

  and

$$\begin{aligned} &\mathbf{v} \in (\mathbf{k}, \vec{\mathcal{H}}_0), \\ &\vec{\mathcal{E}} \perp (\mathbf{k}, \vec{\mathcal{H}}_0), \end{aligned} \tag{2.114}$$

which propagate transverse to the Alfvèn waves and have an opposite circular polarization while propagating along the magnetic field.

For $k_\perp \neq 0$ the magnetohydrodynamic waves are strongly damped. Therefore we can restrict ourselves to the case where the waves propagate along the regular magnetic field $\mathcal{H}_0$, so that $k = k_{||}$, $\mathcal{E}_\perp \neq 0$ and the resonance condition 2.106 gives

$$k_{||} = \pm \frac{\omega_H}{\omega^\alpha(\mathbf{k})/k_{||} - v cos\theta} \approx \pm \frac{Ze\mathcal{H}_0}{pc(\omega^\alpha(\mathbf{k})/kv - \mu)}. \tag{2.115}$$

Here we used the relativistic approximation $E \sim pc$ and $v \sim c$, we introduced $\mu = cos\theta$, where θ is the angle between $\mathbf{p}$ and $\vec{\mathcal{H}}_0$ (so that $v_{||} = vcos\theta = v\mu$). With the above approximations the transport equation 2.100 then becomes

$$\begin{aligned} &\frac{\partial \bar{f}_0}{\partial t} + \mu v \frac{\partial \bar{f}_0}{\partial z} = \\ \pi^2 Z^2 e^2 \sum_\alpha \left(\frac{\omega^\alpha(k)}{kc}\right)^2 &\frac{1}{p^2} \left(\frac{\partial}{\partial p} + \frac{\partial}{\partial \mu}\left(\frac{k_{res} v}{\omega^\alpha(k_{res})} - \mu\right)\right) \times \\ \times \frac{p(1-\mu^2) W^\alpha(k_{res})}{|v\mu - \omega^\alpha(k_{res})/k|} &\left(\frac{\partial}{\partial p} + \left(\frac{k_{res} v}{\omega^\alpha(k_{res})} - \mu\right) \frac{1}{p}\frac{\partial}{\partial \mu}\right) \bar{f}_0, \end{aligned} \tag{2.116}$$

where the following replacements have been performed

$$(p_{||}, p_\perp) \to (p = |\mathbf{p}|, \mu) \Rightarrow \bar{f}_0(t, \mathbf{r}, p_{||}, p_\perp) \to \bar{f}_0(t, \mathbf{r}, p, \mu), \tag{2.117}$$

and we introduced

$$k_{res} = \left|\frac{Ze\mathcal{H}_0}{pc\mu}\right| = \frac{1}{r_H|\mu|}, \tag{2.118}$$

and assumed that the energy density of waves of type α, namely $W^\alpha(k)$, does not depend on phase and polarization. Notice that the wave energy is equally parted between the kinetic energy of the particles of the medium and magnetic field energy

$$\int_0^\infty dk_{||} W^\alpha(k_{||}) = \frac{1}{4\pi}\int_{-\infty}^{\infty} dk_{||} \left|\vec{\mathcal{H}}_1^\alpha(k_{||})\right|, \tag{2.119}$$

by virtue of the approximation $v_A << v$.

2.4.4 Diffusion Approximation

By identifying the term that mostly contributes to 2.116, we can define the effective scattering rate by

$$\begin{aligned}\nu_\mu^\alpha &:= 2\pi^2|\omega_H|\frac{k_{res}W^\alpha(k_{res})}{H_0^2}\left(1-\frac{\omega^\alpha(k_{res})}{k_{res}v}\mu\right)^2 \\ &\approx 2\pi^2|\omega_H|\frac{k_{res}W^\alpha(k_{res})}{H_0^2}.\end{aligned} \tag{2.120}$$

In order to create an isotropic CR flux the relaxation time

$$\tau_{rel}: = (\nu_\mu^\alpha)^{-1} \approx \frac{1}{2\pi^2|\omega_H|}\frac{H_0^2}{k_{res}W^\alpha(k_{res})}, \tag{2.121}$$

is required. Similarly, the relaxation length is defined as $\lambda_{rel}: = v(\nu_\mu^\alpha)^{-1}$. Previously we assumed weak turbulence, i.e. $\nu_\mu^\alpha << |\omega_H|$, and this assumption now acquires the explicit form

$$W^\alpha(k_{res}) << \frac{H_0^2}{2\pi^2 k_{res}}. \tag{2.122}$$

In order to further simplify the propagation equation 2.116, we define two scattering rates corresponding to waves propagating along and opposite to the field H_0

$$\nu_\mu^\pm: \approx 2\pi^2|\omega_H|\frac{k_{res}W^\pm(k_{res})}{H_0^2}, \tag{2.123}$$

where $W^\pm(k)$ are the spectral energies associated to the two propagation directions. Now we can express the transport equation 2.116 as the sum of these two transport modes

$$\begin{aligned}\frac{\partial \bar{f}_0}{\partial t} + \mu v\frac{\partial \bar{f}_0}{\partial z} &= \frac{v_A^2}{p^2}\left(\frac{\partial}{\partial p}p + \frac{v}{v_A}\frac{\partial}{\partial \mu}\right)\frac{1-\mu^2}{2}\nu_\mu^+\frac{p^3}{v^2}\left(\frac{\partial}{\partial p} + \frac{v}{v_A}\frac{1}{p}\frac{\partial}{\partial \mu}\right)\bar{f}_0 + \\ &+\frac{v_A^2}{p^2}\left(\frac{\partial}{\partial p}p - \frac{v}{v_A}\frac{\partial}{\partial \mu}\right)\frac{1-\mu^2}{2}\nu_\mu^-\frac{p^3}{v^2}\left(\frac{\partial}{\partial p} - \frac{v}{v_A}\frac{1}{p}\frac{\partial}{\partial \mu}\right)\bar{f}_0,\end{aligned} \tag{2.124}$$

where we again used $|\omega^\alpha(k)/kv| = v_A/v << 1$. As a consequence of the large value of the factor

$$\left|\frac{k_{res}v}{\omega^\alpha(k_{res})} - \mu\right| \approx \frac{v}{v_A} >> 1, \tag{2.125}$$

that weights the $\partial/\partial\mu$ term in equation 2.116, scattering changes angles rapidly compared to changes in energy. If the anisotropic part $\bar{f}_0$ of the distribution is large enough to satisfy the condition

$$\bar{f}_0 - \hat{f}_0 >> \hat{f}_0 \frac{v_A}{v}, \tag{2.126}$$

where $\hat{f}_0$ is the mean value of $\bar{f}_0$ over the angles

$$\hat{f}_0 = \frac{1}{2} \int_{-1}^{1} d\mu \bar{f}_0, \tag{2.127}$$

then it is reasonable to neglect the energy change of the particle in favor of the angular diffusion. In other words if the time intervals considered are large enough, then the distribution can be assumed almost isotropic. In the following we want to move to the diffusion approximation, i.e. we consider time intervals and distances such that

$$\Delta t >> \tau_{rel}, \ \Delta x >> \lambda_{rel}, \tag{2.128}$$

In this case the expansion

$$\bar{f}_0 = \hat{f}_0 + \delta f(\mu), \ \delta f(\mu) << \hat{f}_0, \tag{2.129}$$

is justified. Substituting 2.129 in 2.124 and keeping only the leading terms ($\delta f << \hat{f}_0$) we get

$$\mu v \frac{\partial \hat{f}_0}{\partial z} = \frac{1}{p^2} \frac{\partial}{\partial \mu} \frac{v_A}{v} \frac{1-\mu^2}{2} (\nu_\mu^+ - \nu_\mu^-) p^3 \frac{\partial \hat{f}_0}{\partial p} + \frac{\partial}{\partial \mu} \frac{1-\mu^2}{2} (\nu_\mu^+ + \nu_\mu^-) \frac{\partial}{\partial \mu} \delta f. \tag{2.130}$$

Averaging over μ yields

$$\frac{\partial}{\partial \mu} \delta f = -\frac{v}{\nu_\mu^+ + \nu_\mu^-} \frac{\partial \hat{f}_0}{\partial z} - \frac{v_A}{v} \frac{\nu_\mu^+ - \nu_\mu^-}{\nu_\mu^+ + \nu_\mu^-} p \frac{\partial \hat{f}_0}{\partial p}, \tag{2.131}$$

which allows to express ∂f in terms of $\hat{f}_0$.

Now we turn again back to 2.124 and insert 2.129, but this time we keep all terms. Averaging over μ gives

$$\frac{\partial \bar{f}_0}{\partial t} + \frac{v}{2}\int_{-1}^{1} d\mu\mu\frac{\partial \delta f}{\partial z} = \frac{v_A^2}{p^2}\frac{\partial}{\partial p}\int_{-1}^{1} d\mu\frac{1-\mu^2}{2}(\nu_\mu^+ + \nu_\mu^-)\frac{p^4}{v^2}\frac{\partial \hat{f}_0}{\partial p} + \\ + \frac{v_A}{p^2}\frac{\partial}{\partial p}p\int_{-1}^{1} d\mu\frac{1-\mu^2}{4}(\nu_\mu^+ - \nu_\mu^-)\frac{p^2}{v}\frac{\partial \delta f}{\partial \mu}. \tag{2.132}$$

Then, by rewriting the second term of the left-hand side as follows

$$\frac{v}{2}\int_{-1}^{1} d\mu\mu\frac{\partial \delta f}{\partial z} = \frac{v}{2}\frac{\partial}{\partial z}\int_{-1}^{1} d\mu\frac{1-\mu^2}{2}\frac{\partial \delta f}{\partial \mu}, \tag{2.133}$$

and using the expression for δf expression that can be derived from 2.131, we end up with

$$\frac{\partial \hat{f}_0}{\partial t} - \frac{\partial}{\partial z}D_{zz}\frac{\partial \hat{f}_0}{\partial z} + \frac{1}{3p^2}\frac{\partial(p^3 u_w)}{\partial p}\frac{\partial \hat{f}_0}{\partial z} - \frac{\partial u_w}{\partial z}\frac{p}{3}\frac{\partial \hat{f}_0}{\partial p} = \frac{1}{p^2}\frac{\partial}{\partial p}p^2 D_{pp}\frac{\partial \hat{f}_0}{\partial p}. \tag{2.134}$$

Here we introduced the effective velocity of convective particle transport by the waves

$$u_w := v_A\int_{0}^{1} d\mu\frac{3(1-\mu^2)}{2}\frac{\nu_\mu^+ - \nu_\mu^-}{\nu_\mu^+ + \nu_\mu^-}, \tag{2.135}$$

which constitutes a drift due to the different energy density $\nu_\mu^\pm$ of the waves along the magnetic field. The spatial diffusion coefficient along the regular magnetic field $\vec{\mathcal{H}}_0$ is defined as

$$D_{zz} = \frac{v^2}{2}\int_{0}^{1} d\mu\frac{1-\mu^2}{\nu_\mu^+ + \nu_\mu^-}, \tag{2.136}$$

and the momentum diffusion coefficient that enters in the stochastic acceleration term is defined by

$$D_{pp} = p^2(\frac{v_A}{v})^2\int_{0}^{1} d\mu 2(1-\mu^2)\frac{\nu_\mu^+\nu_\mu^-}{\nu_\mu^+ + \nu_\mu^-}. \tag{2.137}$$

The particle flux J_z, given by $\frac{\partial \hat{f}_0}{\partial t} = \frac{\partial J_z}{\partial z}$ can be written as the sum of a diffusive and convective term, by just identifying the respective terms in 2.134

$$J_z = \frac{v}{2}\int_{-1}^{1} d\mu\mu\partial f = -D_{zz}\frac{\partial \hat{f}_0}{\partial z} - \frac{u_w}{3}p\frac{\partial \hat{f}_0}{\partial p}. \tag{2.138}$$

Eq. 2.134 is almost the transport equation we were looking for. Except for additional terms due to momentum losses in the ISM, particle losses due to spallation and fragmentation and source terms, this equation forms the basis for all diffusion models.
Two particular cases which greatly simplify equation 2.134 are of special interest

- if the energy density propagating in opposite directions is the same, then $\nu_\mu^+ = \nu_\mu^-$ and the convection velocity 2.135 vanishes;
- if the energy density propagates only in one direction, then the diffusion in momentum space disappears since the coefficient in 2.137 vanishes, i.e. no stochastic acceleration occurs.

Most simple diffusion models, in particular most analytical models, neglect both these terms. The assumption of equal energy density in both transport modes, i.e. $u_w = 0$ is rather arbitrary, but simplifies the transport equation significantly, since only spatial diffusion and diffusion in momentum space remain. This is e.g. the case in the GALPROP code (Strong & Moskalenko, 2006), a widely-used program invoking an isotropic diffusion model which currently is the official Fermi-LAT model for diffuse Galactic γ-rays. On the other hand analytical models usually neglect diffusion in momentum space and only solve the much simpler spatial diffusion equation.

Later in this work, we would like to include the effects of a moving background medium and anisotropic diffusion. Although the motivation to include such processes will be discussed later on (Section 3) we will continue to derive this equation here, because this is only a short way from Eq. 2.134.

2.4.5 A Note on the Terms Convection, Advection, Diffusion and Drift

In the last section we have introduced the term convection describing the drift and subsequent energy losses and gains due to an anisotropy in wave energy density. CRs are transported along with the plasma waves and a resulting energy transport, corresponding to a non-vanishing u_w, results in an additional drift, which we called convection. The term convection has different meanings in different contexts. Especially in the context of CR physics it is not used in a uniform way and is often used synonymous with advection, so a clarification is in order.

Advection is the transport of a substance or conserved property by a fluid in motion. The fluid motion can be described as a vector field, the material or property transported is usually a scalar concentration. An example for advection is e.g. the transport of pollutants (as a substance) or heat (as a property) downstream in a river.

Generally, convection is the movement of molecules within fluids (i.e. liquids, gases and rheids). Convection is one of the major modes of heat and mass transfer. In fluids, convective heat and mass transfer take place through both diffusion and by advection, in which matter or heat is transported by the larger-scale motion of currents in the fluid. In the context of heat and mass transfer, the term "convection" is used to refer to the sum of advective and diffusive transfer. In this general meaning convection comprises CR drift via diffusion (due to a gradient in the source distribution or in the density of scattering centers) or via a general background movement of the ISM.

There is no uniform usage of the terms convection, advection and drift in the literature. In this thesis we will use the term convection to describe both, particle transport via a moving background medium, such as Galactic winds which consist of the material expelled by SNs and can carry CRs with them, and particle transport via a non-vanishing effective wave velocity u_w. We will see in the following, that it is not possible for us to differentiate between these two cases, because both velocities enter the transport equation in the same way. Transport via Galactic winds is a purely advective transport mode, while transport via an effective wave velocity is a mixed advective-diffusive transport mode. It should, however, be noted that *by convection we do not mean particle transport via diffusion*, i.e. the flux generated by one of the two gradients in $\frac{\partial f}{\partial t} = \nabla D \nabla f$. In this case we will refer to the particle transport as "drift via diffusion".

2.4.6 Large-Scale Motion of the Interstellar Medium and Drift

If we want to consider Galactic winds, i.e. gas expelled by SNs moving into the halo with a velocity $|\mathbf{u}(\mathbf{r})| << v$ and whose scale of variation is much larger than the mean path length of the particles, then we have to introduce the large-scale motion of the medium in which the plasma waves propagate. This is easily introduced in Eq. 2.134 with the replacements

$$u_w \rightarrow u_z + u_w, \quad \omega^\alpha \rightarrow \omega^\alpha(k) + k u_z, \tag{2.139}$$

where the velocity of the medium along the regular field is assumed to be

$$\mathbf{u} = \frac{\vec{\mathcal{H}}_0}{|\vec{\mathcal{H}}_0|} u_z. \tag{2.140}$$

We can easily see this by transforming the fields associated with the waves propagating in the medium from the reference system, moving with velocity u_z, to the system at rest. If this transformation is applied to the starting kinetic equation 2.78, then we realize that, up to terms of order v_A/c and u/c, the magnetic field does not change while the electric field becomes

$$\vec{\mathcal{E}}^\alpha = -\frac{1}{c}\left(\frac{\omega^\alpha(k)}{k} + u_z\right)\left(\frac{\vec{\mathcal{H}}_0}{|\vec{\mathcal{H}}_0|} \times \vec{\mathcal{H}}_1^\alpha\right). \tag{2.141}$$

From this equation we can deduce the replacements 2.139. In the following we can perform the steps from equation 2.78 to equation 2.134 and end up with

$$\frac{\partial \hat{f}_0}{\partial t} - \frac{\partial}{\partial z} D_{zz} \frac{\partial \hat{f}_0}{\partial z} + \frac{1}{3p^2}\frac{\partial [p^3(u_z + u_w)]}{\partial p}\frac{\partial \hat{f}_0}{\partial z} - \frac{\partial (u_z + u_w)}{\partial z}\frac{p}{3}\frac{\partial \hat{f}_0}{\partial p} = \frac{1}{p^2}\frac{\partial}{\partial p} p^2 D_{pp} \frac{\partial \hat{f}_0}{\partial p} \tag{2.142}$$

which is identical to Eq. 2.134 except for the replacement $u_w \to u_w + u_z$. The particle flux is then simply replaced by

$$J_z = -D_{zz}\frac{\partial \hat{f}_0}{\partial z} - \frac{u_z + u_w}{3} p \frac{\partial \hat{f}_0}{\partial p}, \tag{2.143}$$

with the contribution of a diffusion term and a convection term, which includes advection via u_z and convection via u_w (see Section 2.4.5 for a definition of convection and advection and their usage in this thesis).

In addition to the motion of the medium along the field, there could be a velocity in arbitrary direction. This is generated by spatial inhomogeneities of the system, by gravitational forces or by the action of the electric field $\vec{\mathcal{E}}_0$. As an example we consider the case of drift generated by a slowly changing electric field

$$\mathbf{u}_\perp = c\frac{\vec{\mathcal{E}}_0 \times \vec{\mathcal{H}}_0}{|\vec{\mathcal{H}}_0|^2}, \tag{2.144}$$

with ($u_\perp << v$) and

$$\vec{\mathcal{E}}_0 = \frac{1}{c}\left(\vec{\mathcal{H}}_0 \times u_\perp\right). \tag{2.145}$$

The electric field does not contribute to the scattering of the particles, but it has to be added to 2.141. We recall the closed differential equation for f_0 Eq. 2.95 and modify it as follows

$$\frac{\partial \tilde{f}_0}{\partial t} + \left(\mathbf{v} \cdot \vec{\nabla}\right) \tilde{f}_0 + \frac{Ze}{c} \left[(\mathbf{v} - \mathbf{u}_\perp) \times \vec{\mathcal{H}}_0\right] \frac{\partial \tilde{f}_0}{\partial \mathbf{p}} =$$

$$= \langle Ze\left(\vec{\mathcal{E}} + \frac{\mathbf{v}}{c} \times \vec{\mathcal{H}}_1\right) \frac{\partial}{\partial \mathbf{p}} \int_{-\infty}^{t} dt' Ze \left(\vec{\mathcal{E}} + \frac{\mathbf{v}}{c} \times \vec{\mathcal{H}}_1\right) \frac{\partial \tilde{f}_0}{\partial \mathbf{p}} \rangle. \tag{2.146}$$

The new term $Ze/c[\mathbf{u}_\perp \times \vec{\mathcal{H}}_0]\partial \tilde{f}_0/\partial \mathbf{p}$ is proportional to $u_\perp$ which is assumed to be much smaller than v, which suggests the expansion

$$\tilde{f}_0 = f_0 + \Delta f, \tag{2.147}$$

where f_0 is the solution of Eq. 2.95 and Δf has to be understood as a small variation induced by the drift due to the electric field variation. Substituting 2.147 in 2.146 and keeping only leading terms, we get

$$\frac{Ze}{c}\left(\mathbf{v} \times \vec{\mathcal{H}}_0\right) \frac{\partial \Delta f}{\partial \mathbf{p}} - \frac{Ze}{c}\left(\mathbf{u}_\perp \times \vec{\mathcal{H}}_0\right) \frac{\partial f_0}{\partial \mathbf{p}} = 0, \tag{2.148}$$

which is solved by

$$\Delta f = -\frac{\mathbf{u}_\perp \cdot \mathbf{p}}{v} \frac{\partial f_0}{\partial \mathbf{p}}, \tag{2.149}$$

The diffusion approximation can now be applied through

$$\tilde{f}_0 = f_0 + \delta f + \Delta f. \tag{2.150}$$

The net effect of this expansion on the propagation equation 2.146 is the introduction of tensorial quantities

$$\frac{\partial f_0}{\partial t} - \nabla_i D_{ij} \nabla_j f_0 + \frac{1}{3p^2} \frac{\partial [p^3(\mathbf{u} + \mathbf{u}_w)]}{\partial p} \cdot (\nabla f_0) - \frac{p}{3} \nabla \cdot (\mathbf{u} + \mathbf{u}_w) \frac{\partial f_0}{\partial p} = \frac{1}{p^2} \frac{\partial}{\partial p} p^2 D_{pp} \frac{\partial f_0}{\partial p}, \tag{2.151}$$

with

$$D_{ij} = D_{||} h_i h_j, \ \mathbf{u}_w = u_w \mathbf{h}, \mathbf{h} = \frac{\vec{\mathcal{H}}_0}{|\vec{\mathcal{H}}_0|}, \tag{2.152}$$

with $D_{\|} = D_{zz}$. This is the full transport equation we were looking for. This equation includes anisotropic and spatially inhomogeneous diffusion by virtue of the diffusion tensor D_{ij}, Galactic winds by virtue of the velocity $\mathbf{u}$ of the background medium, which may take place in any direction and possibly a transport mode due to an anisotropic energy transport by the plasma waves by virtue of $\mathbf{u_w}$. Note, that in the solution of the transport equation we cannot differentiate between $\mathbf{u}$ and $\mathbf{u_w}$, which is why in the following we will attribute the total velocity $\mathbf{u} + \mathbf{u_w}$ entirely to the movement of the background medium. This means that we assume that the magnetohydrodynamic waves propagate equally in both directions and we have ($\mathbf{u}_w = 0$ and $\partial \mathbf{u}/\partial p = 0$)

$$\frac{\partial f_0}{\partial t} - \nabla_i D_{ij} \nabla_j f_0 + (\mathbf{u}\nabla) f_0 - \frac{p}{3} \nabla \cdot \mathbf{u} \frac{\partial f_0}{\partial p} = \frac{1}{p^2} \frac{\partial}{\partial p} p^2 D_{pp} \frac{\partial f_0}{\partial p}. \tag{2.153}$$

This is the one equation which governs all subsequent discussion.

2.4.7 Solar Modulation

We have already seen in Section 2.1.3 that the low-energy spectra of CRs can be modifies by CR transport in our Solar system. The sole reason for this modulation is the interaction of the CR flux from outside the Solar system with the Solar wind which is directed away from the Sun. Depending on the Solar activity this effect can be significant below particle energies of 10-20 GeV. For these energies the LIS predictions of any transport model have to be corrected for the effect of Solar modulation. Here and for the remainder of this thesis we use the so-called force-field approximation, which turns out to be a sufficient approximation in most cases.

Provided, that some specific approximations are taken into account, the propagation equation 2.153 can be applied to the Solar system.

Assuming spherical symmetry in 2.153, only the radial component of the diffusion tensor is left. Furthermore we can safely assume that reacceleration is absent and the Solar wind is radially directed to the outer space. The transport equation 2.153 then reads

$$\frac{\partial f}{\partial t} - \frac{1}{r^2} \frac{\partial}{\partial r} \left(r^2 D_{rr} \frac{\partial}{\partial r} f \right) + u_r \frac{\partial f}{\partial r} - \frac{1}{r^2} \frac{\partial}{\partial r} (r^2 u_r) \frac{p}{3} \frac{\partial f}{\partial p} = 0. \tag{2.154}$$

Gleeson & Axford (1968) showed that it is possible to solve Eq. 2.154 analytically under a number of assumptions. Assuming a spatially constant diffusion coefficient Eq. 2.154

reads in the steady-state case

$$D_{rr}\frac{\partial^2 f}{\partial r^2} + \frac{2D_{rr}}{r}\frac{\partial f}{\partial r} - u_r\frac{\partial f}{\partial r} + \frac{2u_r p}{3r}\frac{\partial f}{\partial p} = 0. \tag{2.155}$$

We can now estimate the importance of the first three terms by considering a very simple diffusion case where the inward diffusion flux is balanced by the outward convection flux

$$D_{rr}\frac{\partial f}{\partial r} = u_r f. \tag{2.156}$$

In this zero approximation case the cosmic rays density is given by the exponential

$$f = F_0 \cdot e^{\frac{u_r}{D_{rr}}r}. \tag{2.157}$$

Substituting the solution 2.157 in 2.155 one sees that the first three terms are of order $u_r^2/D_{rr}f$, $2u_r/rf$ and $u_r^2/D_{rr}f$ respectively, so we have to compare u_r/D_{rr} with $2/r$, i.e. the first and the third term has to be compared to the second term. In this simple model we can write the diffusion coefficient as $D_{rr} = 1/3\lambda v$, with v velocity of the particle. Now we can compare u_r/v to λ/r. During the derivation of the propagation equation we assumed $u_r/v << 1$. Quantitatively we can assume $\lambda \approx 1$ AU and $v \approx c$ for 1 GeV protons and

$$\frac{u_r}{c} \approx 1.3 \cdot 10^{-3} << \frac{\lambda}{r} \approx 0.1 - 10. \tag{2.158}$$

This means that the second term dominates 2.155. The neglection of the first and the third term in 2.155 is called the force-field approximation

$$\frac{\partial f}{\partial r} + \frac{u_r p}{3D_{RR}}\frac{\partial f}{\partial p} = 0, \tag{2.159}$$

which is widely used. With this rather simple expression for CR transport in the Solar system we can focus on the variations in the spectral index of cosmic ray flux. Since all processes that modify the particle density have been excluded, we can consider curves identified by constant f, i.e. curves along $df/dr = 0$. This means we have to solve the system

$$\begin{cases} \frac{df}{dr} = \frac{\partial f}{\partial r} + \frac{dp}{dr}\frac{\partial f}{\partial p} = 0 \\ \frac{\partial f}{\partial r} + \frac{u_r p}{3D_{RR}}\frac{\partial f}{\partial p} = 0 \end{cases}, \tag{2.160}$$

which gives us

$$\frac{dp}{dr} = \frac{u_r p}{3D_{RR}}. \tag{2.161}$$

Assuming that any spatial dependence of the diffusion coefficient can be neglected on the scales of interest we can work in the quasi-linear theory with $\frac{dD_{rr}}{dr} = 0$ and

$$D_{RR} = D_0 \cdot \beta p. \tag{2.162}$$

Inserting this expression in Eq. 2.161 we get

$$\beta dp = \frac{u_r}{3D_0} dr. \tag{2.163}$$

In a final step we introduce the relativistic energy $dE = \beta \cdot dp$ in 2.163

$$dE = \frac{u_r}{3D_0} dr, \tag{2.164}$$

that brings us to the very simple solution

$$E_{ISM} - E(r) = T_{ISM} - T(r) = \Phi \frac{R_{ISM} - r}{R_{ISM} - 1}, \tag{2.165}$$

where $T(r)$ is the kinetic energy and Φ is called modulation strength. The subscript ISM stands for the distance between the Earth and the interstellar medium and the factor $R-1$ has been introduced to give to the modulation strength

$$\Phi = \frac{u_r (R_{ISM} - 1)}{3D_0}, \tag{2.166}$$

the meaning of the energy loss between the ISM and 1 AU as can be seen by setting $r = 1$ AU in Eq. 2.165

$$\Phi = T_{ISM} - T(1 \text{ AU}). \tag{2.167}$$

The units used here are MeV for the energy and AU for the distance. Consequently the modulation strength given here is in MV. Equation 2.166 is widely used because of its simplicity and usability. Nevertheless one has to keep in mind that we made a number of simplifying assumptions to get here and in the end enclose the entire Solar modulation problem, i.e. all processes that occur between the detector and the heliopause, in just one single parameter Φ. This modulation parameter is model-dependent, it does not represent the Solar potential itself. A value of Φ can only be quoted in the context of a given propagation model. Note, that the force field approximation above is independent of the charge of the sign of the CR's charge. It has been suggested in the literature for a long time that this approximation might not hold. The recent measurements of the PAMELA experiment on electrons and positrons as well as protons and antiprotons seem to indicate that there is indeed a charge sign dependent Solar modulation (Gast & Schael, 2009).

Chapter 3

Models for Cosmic Ray Transport

While most objects in the Milky Way can only be observed indirectly via the electromagnetic radiation produced in some process, CRs can be directly sampled which makes them a unique probe of Galactic astrophysics. Other examples are meteorites and stardust. CRs provide us with a detailed elemental and isotopic sample of the current (few million years old) interstellar medium not available in any other way. It is this which makes the subject especially rich and complementary to other disciplines. A detailed CR transport model can help us to understand the physical processes in the CR acceleration regions, the most energetic regions in our Galaxy. It can be used as a cross check of our models of the local ISM and to smaller extend of the ISRF and the magnetic field. For most indirect dark matter searches in diffuse γ-rays, synchrotron radiation or charged annihilation or decay products Galactic CRs form the dominant background. Practically all our knowledge of CR propagation comes via secondary CRs, with additional information from γ-rays and synchrotron radiation. The fact that the primary nuclei are measured (at least locally) means that the secondary production functions can be computed from primary spectra, cross sections and interstellar gas densities with reasonable precision; the secondaries can then be "propagated" and compared with observations. Since the realization that CR fill the Galaxy it has been clear that nuclear interactions imply that their composition contains information on their propagation (Bradt & Peters, 1950). A historical event was the arrival of satellite measurements of isotopic *Li*, *Be*, *B* in the 1970's (Garcia-Munoz et al., 1975). Since then the subject has expanded enormously with models of increasing degrees of sophistication. The simple observation that the observed composition of CR is different from the Solar abundances, in that rare Solar system nuclei like Boron are abundant in CR, proves the importance of propagation in the interstellar medium. At present we believe that the diffusion model with possible inclusion of convection provides the most adequate description of CR transport in the Galaxy at energies below about 10^{17} eV so we begin by presenting this model.

In this chapter, we introduce the GALPROP code, which utilizes an isotropic diffusion model. After a detailed description of our model of the Milky Way in terms of gas distribution, ISRF, magnetic field and CR sources, we will discuss some of the isotropic diffusion models in the GALPROP frame, which constitute the state-of-the-art in CR transport modelling. We will end this chapter with a comparison of these models to recent observations and well established knowledge about our Galaxy and show that the isotropic GALPROP models are incompatible with some of our observations and that there is evidence for anisotropic CR transport.

3.1 Isotropic Propagation Models

The main features of our Galaxy are a barred central bulge with a diameter of a few kpc and a large spiral disk with a radius of about 15 kpc and an density falling exponentially in R with a scale length of about 2.5 kpc and in z with a scale height of about 0.25 kpc. Most of the gas is distributed in the disk with a broad maximum between $R = 4$ kpc and 6 kpc for molecular hydrogen, while the distribution of ionized hydrogen is nearly constant between $R = 4$ kpc and 13 kpc. The supernovae remnants (SNR) are also distributed in the disk, but peak at a distance of a few kpc from the center with a slow fall off to larger radii (Case & Bhattacharya, 1996), as will be shown later. The CRs form a plasma of ionized particles, in which the electric fields can be neglected by virtue of the high conductivity and the magnetic fields form Alfvén waves, where the ion mass density provides the inertia and the magnetic field line tension provides the restoring force. If the wavelength of the Alfvén waves equals a multiple of a particle gyroradius, resonant scattering occurs, leading to a change in the CRs pitch angle without energy losses. Such a process leads to a random walk, which can be described by a diffusion equation as derived in Section 2.4.6

$$\frac{\partial f}{\partial t} = \nabla_i D_{ij} \nabla_j f - (\vec{u}\nabla) f + \frac{p}{3} \nabla \cdot \vec{u} \frac{\partial f}{\partial p} + \frac{1}{p^2} \frac{\partial}{\partial p} p^2 D_{pp} \frac{\partial f}{\partial p}, \tag{3.1}$$

where $f(\vec{r}, \vec{p}, t)$ is the CR phase space density, D_{ij} are the components of the diffusion tensor for spatial diffusion, D_{pp} the diffusion coefficient in momentum space and $\vec{u}$ is the convection velocity. Convective transport is possible either by a large scale motion of the

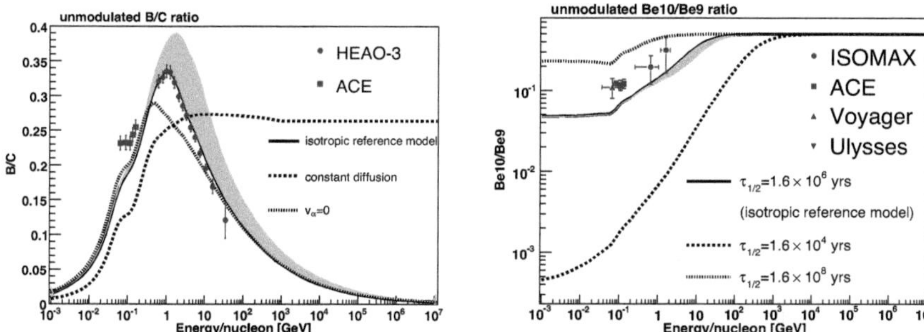

Figure 3.1: Left Demodulated local B/C ratio for different variations of an isotropic model: isotropic reference model - a model similar to the conventional GALPROP model (Strong et al., 2004a)- (full line), isotropic model with constant (energy independent) diffusion (dashed line), isotropic model without diffusive reacceleration (dotted line). For energy independent diffusion (which means the escape time does not depend on energy), the B/C ratio is independent of energy once the nuclei become relativistic. Setting the Alfvén velocity v_α to zero eliminates diffusive reacceleration, which therefore shifts the peak in the B/C ratio to lower energies. The grey band refers to a run with increased halo height (from $z_h = 4$ kpc to $z_h = 5.3$ kpc). *Data:* HEAO-3 (Engelmann et al., 1990) ACE (Davis, 2000) **Right** Demodulated local $^{10}Be/^{9}Be$ ratio for different lifetimes of ^{10}Be: $\tau_{1/2} = 1.6 \cdot 10^6$ yrs (full line), $\tau_{1/2} = 1.6 \cdot 10^4$ yrs (dashed line), $\tau_{1/2} = 1.6 \cdot 10^8$ yrs (dotted line). The grey band refers to a run with $z_h = 5.3$ kpc. *Data:* ISOMAX (Hams et al., 2004), ACE (Yanasak et al., 2001), Voyager (Lukasiak, 1999), Ulysses (Connell et al., 1998).

interstellar medium with velocity $\vec{V}$ or by the effective velocity of the Alfvén waves $\vec{u}_w$. Assuming that the energy density in Alfvén waves propagating in opposite directions is the same $\vec{u}_w$ vanishes and $\vec{u} = \vec{V}$. In addition to spatial diffusion, diffusion in momentum space can be caused by CRs scattering on moving Alfvén waves, which leads to diffusive reacceleration. One finds the corresponding diffusion coefficient to be $D_{pp} \propto p^2 v_\alpha^2 / D_{zz}$, where the Alfvén velocity v_α is introduced as a characteristic velocity of weak disturbances propagating in a magnetic field (Strong et al., 2007). The Alfvén waves propagate in the direction of the magnetic field with velocity $v_\alpha = B/\sqrt{4\pi\rho}$, where B is the magnetic field and ρ the plasma density. Waves at oblique incidence exist and Alfvén waves smoothly change into magnetosonic waves when the propagation is perpendicular to the magnetic field. It is usually assumed that the steady state condition is reached, i.e. $\partial f/\partial t = 0$, which implies that the injection rate of CRs by sources equals the loss rate. CRs can be

lost either by fragmentation, decay or escape from the Galaxy. In the steady state case the diffusion equation for CRs can be solved numerically for given boundary conditions. Usually one requires the CR density to become zero above a certain halo height.
Most primary nuclei show a power law spectrum falling with rigidity like $E^{-2.54}$. This can be easily reproduced by selecting the injection spectrum of the primary particles, the rigidity dependence of the diffusion coefficient and the energy gains due to diffusive reacceleration accordingly. During their journey CRs may interact with e.g. the gas in the Galaxy and produce secondary particles. This changes the ratio of secondary/primary particles, like the B/C ratio. From this ratio one can determine the grammage, which was found to be of the order of 10 g/cm^{-2} (Ginzburg et al., 1990; Schlickeiser, 2002). This corresponds to a density of about 0.2 atoms/cm^3, which is significantly lower than the averaged density of the disk of about 1 atom/cm^{-3}. Under the assumption of a homogeneous gas distribution this suggests that CRs travel a large time in low density regions, like the halo.
For relativistic energies the inelastic cross sections for secondary particle production usually do not strongly depend on the energy of the particle. This would lead to rather flat spectra for the secondary/primary ratios in contrast to the observed B/C ratio, which shows a maximum at about 1 GeV/nucleon and decreases as $E^{-0.6}$ towards higher energy, as shown by the dashed line in the right hand side of Fig. 3.1. This can be accommodated by assuming that energetic particles diffuse faster out of the Galaxy, i.e. the energy dependence of the diffusion coefficient and the energy gains due to diffusive reacceleration are chosen accordingly (see the full line in the left hand side of Fig.3.1). The decrease at low energies can be accommodated by both diffusive reacceleration, which shifts the spectrum to higher energies, as well as convective transport mechanisms (Strong et al., 2007). Alternatively, one could assume a strong increase of the diffusion coefficient at low energies due to damping of the Alfvén waves (Ptuskin et al., 2006).
From the ratio of unstable/stable secondary nuclei (like $^{10}Be/^{9}Be$) one obtains the average residence time of CRs in the Galaxy to be of the order of $t_{CR} = 10^7$ yrs (Cesarsky, 1980). Fig. 3.1 shows the local $^{10}Be/^{9}Be$-fraction assuming different lifetimes of ^{10}Be. The data require residence times between $1.6 \cdot 10^6$ and $1.6 \cdot 10^8$ years.

3.2 The GALPROP Code

A widely-used program providing a numerical solution to the diffusion equation is the publicly available GALPROP code (Strong & Moskalenko, 2006). CR transport in GALPROP is completely based on the kinetic theory discussed in Chapter 2. The code numerically solves the transport equation including Galactic winds (convection), diffusive reacceleration in the interstellar medium, energy losses, nuclear fragmentation and decay with a given source distribution and user-defined boundary conditions (assuming free escape of CRs beyond the boundary) for all CR species. The numerical solution of the transport equation is based on a Crank-Nicholson (Press et al., 1992) implicit second-order scheme. Since we have a 3-dimensional (R, z, p) or 4-dimensional (x, y, z, p) problem (spatial variables plus momentum) one uses "operator splitting" to handle the implicit solution. The transport equation in GALPROP is not written in the form of a phase space density $f(\mathbf{r}, \mathbf{p}, t)$, but in the form of density per unit of total particle momentum $\Psi(\mathbf{r}, p, t)$ defined by

$$\Psi(p)dp = 4\pi p^2 f(\mathbf{p})dp, \tag{3.2}$$

since this is the natural unit for propagation. With the convection velocity $\mathbf{u} = \mathbf{V}$, the forth term in Eq.2.153 then can be rewritten as follows

$$\frac{1}{3}p(\vec{\nabla}\cdot\mathbf{u})\frac{\partial}{\partial p}\left(\frac{\Psi}{p^2}\right) = \frac{1}{3}p(\vec{\nabla}\cdot\mathbf{V})\frac{\partial}{\partial p}\left(\frac{1}{p^3}p\cdot\Psi\right) = -\frac{1}{p^2}(\vec{\nabla}\cdot\mathbf{V})\Psi + \frac{1}{3p^2}(\vec{\nabla}\cdot\mathbf{V})\frac{\partial}{\partial p}(p\Psi). \tag{3.3}$$

Using 3.3 and 3.2 and multiplying the transport equation 2.153 by p^2 one obtains

$$\frac{\partial\Psi}{\partial t} = \vec{\nabla}\cdot(D\vec{\nabla}\Psi - \vec{V}\Psi) + \frac{\partial}{\partial p}p^2 D_{pp}\frac{\partial}{\partial p}\frac{1}{p^2}\Psi + \frac{1}{3}\frac{\partial}{\partial p}\left[\left(\vec{\nabla}\cdot\mathbf{V}\right)p\Psi\right], \tag{3.4}$$

where it is assumed that the diffusion coefficient D is a scalar quantity with the same value everywhere and in all directions, i.e. the tensor D_{ij} has only diagonal components, which are all equal. The full transport equation used in GALPROP can then be written as

$$\frac{\partial\Psi}{\partial t} = q(\vec{r},t) + \vec{\nabla}\cdot(D\vec{\nabla}\Psi - \mathbf{V}\Psi) + \frac{\partial}{\partial p}p^2 D_{pp}\frac{\partial}{\partial p}\frac{1}{p^2}\Psi - \frac{\partial}{\partial p}\left[\dot{p}\Psi - \frac{p}{3}\left(\vec{\nabla}\cdot\mathbf{V}\right)\Psi\right] - \frac{1}{\tau_f}\Psi - \frac{1}{\tau_r}\Psi \tag{3.5}$$

where $q(\vec{r},t)$ is the source term, τ_f is the time scale for fragmentation and τ_r is the time scale for radioactive decay.

The basic parameters for GALPROP are the injection spectrum parameters, the diffusion coefficient D, the convection velocity $\mathbf{V}$, Alfvén velocity v_α, which enters D_{pp}, and the size of the transport box, needed as a boundary to solve the differential equation. Usually one assumes the density of scattering centers outside the diffusion box to be so small that CRs propagate freely with the speed of light to outer space (free escape). Consequently the CR

density becomes zero at the boundary.
GALPROP employs the so-called "test-particle approach", meaning that all transport parameters are treated as fixed quantities while the CR density assumes a stationary state with respect to those fixed parameters. Specifically this means that the spectrum and the amplitude of turbulence as described by the diffusion coefficient and v_α does not depend on the evolving CR distribution itself, although this distribution is expected to generate the turbulence in reality. Since with GALPROP we are looking for a steady-state solution and we can safely assume that the fraction of "fresh" CRs is generally small, the test-particle approach is a reasonable approximation, because it forces the solution to adapt to a fixed spectrum of turbulence, which can be considered to be generated by the steady-state solution we are looking for. Following Strong & Moskalenko (2006) we briefly describe the numerical procedure applied in GALPROP.

Numerical Solution of the Transport Equation The diffusion, reacceleration, convection and loss terms in Eq. 3.5 can be finite-differenced for each dimension (R, z, p) in the form

$$\frac{\partial \Psi_i}{\partial t} = \frac{\Psi_i^{t+\Delta t} - \Psi_i^t}{\Delta t} = \frac{\alpha_1 \Psi_{i-1}^{t+\Delta t} - \alpha_2 \Psi_i^{t+\Delta t} + \alpha_3 \Psi_{i+1}^{t+\Delta t}}{\Delta t} + q_i, \tag{3.6}$$

where all terms are functions of (R, z, p) and the index i indicates the type of nucleus under consideration. In the Crank-Nicholson implicit method (Press et al., 1992) the updating scheme is

$$\Psi_i^{t+\Delta t} = \Psi_i^t + \alpha_1 \Psi_{i-1}^{t+\Delta t} - \alpha_2 \Psi_i^{t+\Delta t} + \alpha_3 \Psi_{i+1}^{t+\Delta t} + q_i \Delta t. \tag{3.7}$$

The tridiagonal system of equations,

$$-\alpha_1 \Psi_{i-1}^{t+\Delta t} + (1 + \alpha_2) \Psi_i^{t+\Delta t} - \alpha_3 \Psi_{i+1}^{t+\Delta t} = \Psi_i^t + q_i \Delta t, \tag{3.8}$$

is solved for the $\Psi_i^{t+\Delta t}$ by the standard method (Press et al., 1992). The three spatial boundary conditions

$$\Psi(R, z_h, p) = \Psi(R, -z_h, p) = \Psi(R_h, z, p) = 0 \tag{3.9}$$

are imposed at each iteration. No boundary conditions are imposed or required at $R = 0$ or in p. Grid intervals are typically $\Delta R = 1$ kpc, $\Delta z = 0.1$ kpc; for p a logarithmic scale with ratio typically 1.2 between successive energies is used. Although a finer grid would allow to model small scale phenomena such as density variations in the ISM, the maximum resolution is tightly limited by the available memory capacities [1]. The model is symmetric around $z = 0$, but the solution is generated for $-z_h < z < z_h$ since this is required for the tridiagonal system to be valid. For the 3-dimensional (R, z, p) problem we handle the implicit solution as follows. One applies the implicit updating scheme alternately for the operator in each dimension in turn, keeping the other two coordinates fixed. To account for the sub steps $1/3q_i$ and $1/3\tau$ are used instead of q_i and τ. The method was found to be stable for all α (Strong & Moskalenko, 2006), and this property can be exploited to advantage by starting with $\alpha >> 1$ (see below). The standard alternating direction implicit (ADI) method, in which the full operator is used to update each dimension implicitly in turn, is more accurate but was found to be unstable for $\alpha > 1$. This is a disadvantage when treating problems with many timescales, but can be used to generate an accurate solution from an approximation generated by the non-ADI method. A check for convergence is performed by computing the timescale $\frac{\Psi}{\partial\Psi/\partial t}$ and requiring that this be large compared to all diffusive and energy loss timescales. The main problem in applying the method in practice is the wide range of time-scales, especially for the electron case, ranging from 10^4 years for energy losses to 10^9 years for diffusion around 1 GeV in a large halo. Use of a time step Δt appropriate to the smallest time-scales guarantees a reliable solution, but requires a prohibitively large number of steps to reach the long time-scales. The following technique was found to work well: start with a large Δt appropriate for the longest scales, and iterate until a stable solution is obtained. This solution is then accurate only for cells with $\alpha << 1$; for other cells the solution is stable but inaccurate. Then reduce Δt by a factor (0.5 was adopted) and continue the solution. This process is repeated until $\alpha << 1$ for all cells, when the solution is accurate everywhere. It is found that the inaccurate parts of the solution quickly decay as soon as the condition $\alpha < 1$ is reached for a cell. As soon as all cells satisfy $\alpha < 1$ the solution is continued with the ADI method to obtain maximum accuracy. A typical run starts with $\Delta t = 10^9$ years and ends with $\Delta t = 10^4$ years for nucleons and 10^2 years for electrons performing ~60 iterations per Δt. In this way it is possible to obtain reliable solutions with reasonable computer resources, although

[1]For a model with spatially dependent transport parameters as introduced in Chapter 4 and run over all nuclei from $Z = 28$ to $Z = 1$ plus the additional CR species from dark matter annihilation (see Chapter 5) between 8 and 16 GB memory are required. For simpler isotropic models with constant transport parameters and no contribution from dark matter annihilation the memory requirements can be reduced to 2-4 GB for a grid spacing as given in the text.

the CPU time required is still considerable. All results are output as FITS[2] datasets for subsequent analysis.

Since lighter nuclei can have source terms from the spallation or decay of heavier nuclei, the propagation equation is solved first for the heaviest nuclei (A) and all secondary source functions are calculated from the steady state solution. The production of secondary and tertiary particles is calculated in GALPROP using a network with more than 2000 cross sections. This step is then repeated for the next lighter nuclei ($A-1$) up to hydrogen ($A=1$). Finally the electrons and positrons are propagated. With this algorithm all secondaries and tertiaries are automatically taken into account. The whole chain is usually repeated twice to increase the accuracy of the predicted secondary to primary ratios and of $\beta^{\pm}$-instable isotopes.

Internal Units The kinetic energy of nuclei is internally given as kinetic energy per nucleon E_{kin}. The reasoning behind this, is that the secondary-to-primary computation is simplified since in this case primaries produce secondaries of the same E_{kin}. On the other hand the basic CR density used has units of density per total momentum p since this is the natural unit for propagation. The combined requirements of transport and secondary production can be elegantly met if one uses $\frac{c}{4\pi}n(p)$ internally, where $n(p) = dn/dp$ in units of $\mathrm{cm^{-3}MeV^{-1}}$. In this case the flux $I(E_{kin})$ in $\mathrm{cm^{-2}sr^{-1}s^{-1}(MeV/nucleon)^{-1}}$, which is useful for comparison with observations is simply given by

$$I(E_{kin}) = \frac{\beta c}{4\pi}\frac{dn}{dp}\frac{dp}{dE_{kin}} = \frac{c}{4\pi}n(p)A, \qquad (3.10)$$

where A is the mass number of the nucleus (or 1 in the case of electrons and positrons), and E_{kin} corresponds to the total kinetic energy). Equation 3.10 follows from $dp = \frac{A}{\beta}dE_{kin}$.

3.3 The Milky Way Model

Any model for CR transport is based on two fundamental types of simplification, which outline the edge of our knowledge in the respective fields: On the one hand the transport

[2]FITS stands for "Flexible Image Transport System" and is the standard astronomical data format endorsed by both NASA and the IAU (International Astronomical Union). FITS is primarily designed to store scientific data sets consisting of multi-dimensional arrays (1-D spectra, 2-D images or 3-D data cubes) and 2-dimensional tables containing rows and columns of data.

equation already comprises assumptions about the relevant physical processes by means of the approximations that were made in deriving the equation or the neglection of certain processes. On the other hand the model of our Milky Way, its gas content, magnetic fields etc., constitute a significant uncertainty since the exact properties of our Galaxy are generally poorly known and have to be derived very carefully from the column density measurements.
Here we will briefly describe the GALPROP model of the Milky Way, which - in the field of CR transport models - is currently the most detailed description of our Galaxy.

GALPROP is designed to treat both two and three spatial dimensional models. In order to reduce the CPU and memory requirements we chose a cylindrical symmetry. In this framework the Galaxy is considered as a dense central disk of thickness $2h$ where h is around 100 pc, surrounded by a cylindrical halo where cosmic rays are still trapped by the Galactic magnetic field. The gas content and CR sources are located in the central disk. The half height of the halo is one of the most important parameters, usually running between a few kpc to $\sim$ 20 kpc as suggested by previous studies on radioactive nuclei (Lukasiak, 1994) and distribution of synchrotron radiation (Phyllips, 1981). The radial extension of the box is fixed to $20 - 30$ kpc, which means it extends beyond the gaseous disk ($\sim$ 15 kpc). Outside of the transport box free escape of cosmic rays is assumed, i.e. the CR density is set to zero. The Solar system is located at $R = 8.3$ kpc and $z = 0$.

The structure of the Galaxy is included in the form of the gas content, which is important for secondary production and energy losses, the interstellar radiation field (ISRF) and magnetic field, which both strongly affect electron energy losses and lead to the production of γ-rays from IC and synchrotron radiation. The distribution of atomic hydrogen is reasonably well known from 21-cm surveys, but the distribution of molecular hydrogen can only be estimated using CO as a tracer. The Galactic magnetic field is best determined from Faraday rotation measurements from the polarized radio emission of pulsars with known distances combined with a model for the distribution of ionized gas. The ISRF consists of contribution from cosmic microwave background and starlight and is modified by absorption and reemission by interstellar dust.

3.3.1 The Interstellar Gas

The space between the stars in our Milky Way is populated by diffuse matter. The interstellar medium mainly consist of hydrogen (70%) followed by helium (28%) and only a 2% contribution from heavier elements. Hydrogen is present in three possible forms: atomic hydrogen HI, molecular hydrogen H_2 and ionized hydrogen HII.

Atomic hydrogen occurs in cold clouds of temperature $T \simeq 50$ K and density $n \simeq 30$ cm^{-3} or in the form of diffuse matter with temperature $T \simeq 10^4$ K and density $n \simeq 0.1$ cm^{-3}.

Averaged over the azimuthal coordinate Φ and radial sizes of 0.5 kpc a good fit to the atomic hydrogen distribution is given by an exponentially decreasing function of the halo height (Burton, 1988)

$$n_{HI}(r,z) = n^0_{HI}(r) e^{-\frac{z^2}{2h^2_{HI}(r)}} \mathrm{cm}^{-3}, \tag{3.11}$$

where $n_{HI}(r)$ is taken from Gordon & Burton (1976) and shown in figure 3.2 while $h_{HI}(r)$ has the form proposed in Cox et al. (1986), namely

$$h_{HI}(r) = \begin{cases} 0.25 \text{ kpc}, & r \leq 10\text{kpc} \\ 0.083 e^{0.11r} \text{ kpc}, & r > 10\text{kpc}. \end{cases} \tag{3.12}$$

Molecular hydrogen can be traced with the $\lambda = 2.6$ mm $(J = 1 \rightarrow 0)$ emission line of CO, since collisions between the CO and H_2 molecules in the clouds are responsible for the excitation of CO. The large-scale density distribution can be represented by (Bronfman, 1988)

$$n_{H_2}(r,z) = n^0_{H_2}(r) e^{-\frac{z^2}{2h^2_{H_2}}} \mathrm{cm}^{-3}, \tag{3.13}$$

with $h_{H_2} = 0.05 \pm 0.01$ kpc. The averaged radial dependence of n_{H_2} is again reported in figure 3.2, it shows a pronounced peak between 4 and 6 kpc and has values between 0.15 an 0.45 H_2 molecules cm^{-3} in the inner Galaxy $(R < R_\odot)$ whereas in the outer Galaxy the density is very small. H_2 can exist only in dark clouds where it is protected from the ionizing UV radiation from stars. Roughly 40% of the mass of the ISM occurs in this form. However, the volume filling factor of these clouds is below 2%, which means that the number density of the clouds must be of the order 10^3 molecules cm^{-3}. Given the fact that these clouds are accompanied by large magnetic fields this could have a significant impact on CR propagation and secondary production: CRs can be reflected from the high magnetic fields in the clouds, thus reducing the number of CR interactions in the H_2 component of the ISM significantly. Due to their strong frozen-in magnetic fields molecular clouds align in the form of cloud complexes, where the single clouds are interconnected by weak magnetic fields. The molecular clouds in one complex then form magnetic mirrors which could repeatedly reflect and thus trap CRs for a significant amount of time in the intercloud material of the cloud complex, excluding them from the molecular clouds themselves. These processes occur on very small length scales, likely below the mean scattering length of diffusion (Chandran, 2000). Trapping by molecular cloud magnetic mirrors will be discussed in Chapter 5 in more detail. For now we will assume that CRs propagate in H_2 as in the rest of the ISM, so that the clumpiness of the H_2 component has no influence on CR propagation and secondary production. Consequently, the H_2 distribution can be modelled by the averaged gas density.

Ionized hydrogen occurs in the vicinity of young O and B stars. Here the UV radiation from the stars ionizes the ISM. HII regions are distributed similar to the molecular hydrogen, but mass-wise their contribution is negligible. Since they only occur close to bright stars, HII regions are a good tracer of the spiral structure of the Milky Way. The averaged large scale distribution of this gas components can be parameterized as (Cordes, 1991)

$$n_{II}(r,z) = \left(0.025e^{-\frac{|z|}{1\mathrm{kpc}}-\left(\frac{r}{20\mathrm{kpc}}\right)^2} + 0.2e^{-\frac{|z|}{0.15\mathrm{kpc}}-\left(\frac{r}{2\mathrm{kpc}}-2\right)^2}\right)\mathrm{cm}^{-3}, \tag{3.14}$$

where the second term takes the concentration around r=4 kpc into account.

Other components of the ISM The so-called *coronal gas component* is negligible in mass, but it may occupy up to 80 % of the volume of the interstellar gas. This gas exists in the form of a very dilute (density $n \simeq 10^{-3}$ cm^{-3}) hot ($T \simeq 10^5 - 10^6$ K) plasma and is very similar to the solar corona. This component can be traced by soft X-rays and ultraviolet OVI absorption lines. According to our current understanding the coronal component results from intersecting SNR, where the outgoing SN shock sweeps up interstellar material and leaves behind a dilute hot gas (Schlickeiser, 2002).

Finally, *helium* can be traced by photospheric methods (Grevesse et al., 1996). Helium appears to follow the hydrogen distribution with a factor $He/H = 0.10 \pm 0.08$. We adopt a value of $He/H = 0.11$, which is widely used in the literature. The uncertainties in secondary production due to a possibly smaller value are negligible and much smaller than the uncertainties of the hydrogen distribution itself.

Table 3.1 shows an overview over the different phases of the ISM and the respective radio tracers.

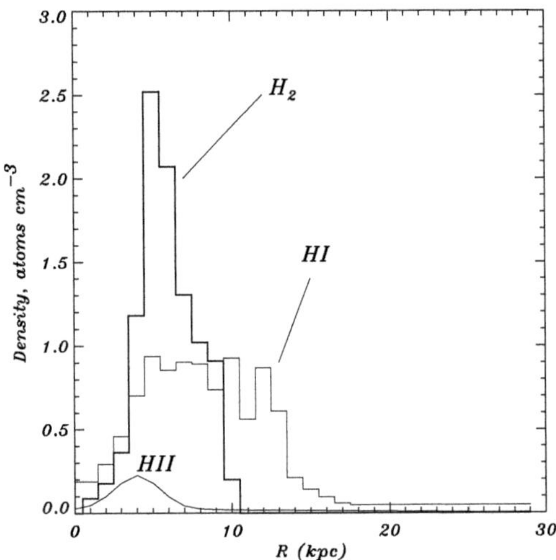

Figure 3.2: Schematic profile of the radial dependence of the three components of hydrogen gas as a function of radius at z=0 kpc from Strong & Moskalenko (1998).

3.3.2 Source Distribution

In Section 2.1.2 we argued that supernovae are a reasonable source of cosmic rays. Assuming to have a supernova event every 10^4 yrs inside a volume of a cubic kpc and assuming that an SNR accelerates CRs for 10^4 yr we have at least one cosmic ray acceleration site every cubic kpc at any time. The derivation of the Galactic distribution of SNRs, commonly based on radio surveys as in the case of Case & Bhattacharya (1996), is subject to large observational selection effects. Other tracers of the SNR distribution are available, in particular pulsars. The Parkes Multibeam survey with 914 pulsars has been used by Lorimer (2004) to derive the SNR distribution, which basically confirms the concentration of SNR in the inner Galaxy. The SNR distribution can be parameterized by (Case & Bhattacharya, 1996)

$$Q(R, z) = q_0 \left(\frac{R}{R_0}\right)^{\alpha} e^{-\beta\frac{(R-R_0)}{R_0}} e^{\frac{-|z|}{z_s}}, \tag{3.15}$$

where q_0 is a normalization constant, R_0 is the Galactocentric radius of the Sun, $\alpha = 1.69$, $\beta = 3.22$ and $z_s = 0.2$ kpc.

State	Gas Phase	Number density [cm^{-3}]	Kinetic temperature [K]	Volume filling [%]	Mass fraction [%]	Radio tracers
Molecular	Giant molecular clouds	10^3	10	≤ 2	40	CO, NH_3, H_2CO
	Dark clouds	$10^2 - 10^3$	10			
Atomic	HI clouds	30	50-100		40	HI
	Intercloud	$0.1 - 1$	$10^3 - 10^4$	50	20	
Ionized	HII regions	$1 - 10^5$	10^4	≤ 2	≤ 1	Recomb. Lines Free-Free
	Coronal gas	$10^{-4} - 10^{-2}$	$10^4 - 10^6$	$20 - 80$	≤ 1	UV-absorp. lines soft X-rays

Table 3.1: Phases of interstellar matter (after Downes & Guesten (1982))

Figure 3.3 shows the SNR distribution and the pulsar distribution from Lorimer (2004). Also shown is the flattened source parameterization by Strong & Moskalenko (1998) as inferred from the soft γ-ray gradient in the COS-B and EGRET data.

It has been suggested that a considerable amount of C and O is accelerated in the wind material of C- and O-enriched pre-supernovae Wolf-Rayet stars (Meyer et al., 1997). Since these stars generally coincide with the SNR distribution this does not affect the source model.

3.3.3 Injection Spectrum

After the initial acceleration process in the SN shock fronts CRs are injected into the ISM and their transport is governed by the transport equation 3.5. At the end of Section 2.3.2 we found that a power law in momentum

$$q(E) = q_0 Q(E_{kin}), \ \frac{dQ(E_{kin})}{dE_{kin}} \propto p^{-\gamma}, \tag{3.16}$$

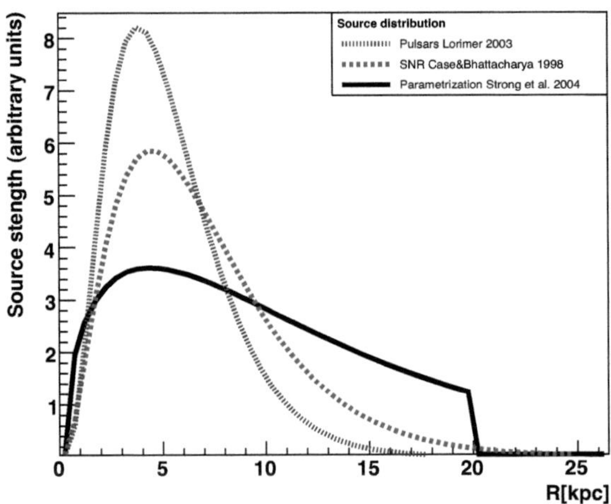

Figure 3.3: Source distribution: pulsar distribution (Lorimer, 2004) (*blue, dotted*), SNR distribution (Case & Bhattacharya, 1996) (*red, dashed*) and the flattened parameterization used by Strong & Moskalenko (1998) (*black full*).

would best describe the initial energy spectrum. Here q_0 is a normalization constant and $Q(E_{kin})$ is the source flux. Remember that Eq. 3.10 tells us that we are already in the proper units for the transport equation used in GALPROP. Since the rigidity $\rho = p/Z$ is the property that governs the CR transport equation it is preferable to write Eq. 3.16 in the following form

$$\frac{dI_s(E_{kin})}{d\rho} \sim \left(\frac{\rho}{\rho_0}\right)^{-\gamma}, \tag{3.17}$$

to account for a possible break at a reference rigidity ρ_0 with different values of γ above and below ρ_0.

3.3.4 Interstellar Radiation Field

For the calculation of the spectrum of γ-rays arising from inverse Compton scattering and electron energy losses, the full ISRF as function of (R, z, ν) is required. The ISRF consists of contributions from starlight, emission from dust, and the CMB. The estimation of the spectral and spatial distribution of the ISRF therefore crucially relies on models of

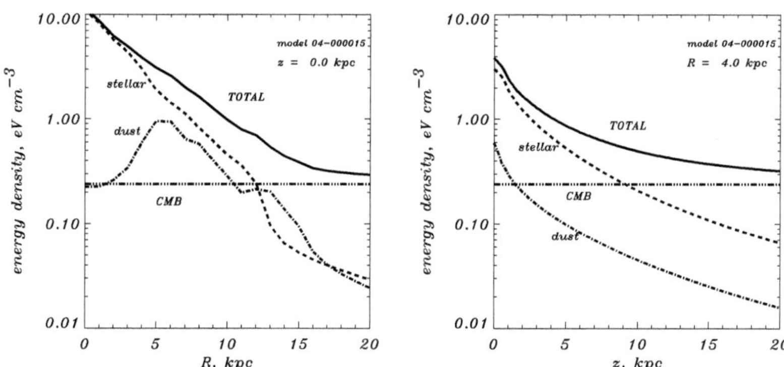

Figure 3.4: ISRF energy density as a function of R at z = 0 (**left**), and of z at R = 4 kpc (**right**). Shown are the contributions of stars (dashed), dust (dash-dot), CMB (dash-3-dots), and total (full line). From Strong et al. (2000)

the distribution of stars, absorption, dust emission spectra and emissivities. GALPROP uses the ISRF calculation provided by Moskalenko et al. (2006); Porter (2005); Strong & Moskalenko (2006). Fig. 3.4 shows the spatial distribution and composition of the ISRF as used in GALPROP.

3.3.5 Galactic Magnetic Field

Although the uniform magnetic field is a key parameter for CR transport, since it influences the CR scattering rate (together with the perturbed component of the field) as well as the size of the CR confinement volume, i.e. the halo height, the magnetic field does not enter the CR transport equation directly. The reason for this is, that in the test particle approach we are using here, all properties of the magnetic field are "hidden" in the diffusion coefficient and the Alvén velocity.

In GALPROP the Galactic magnetic field is only used for the calculation of synchrotron losses for electrons and positrons. Investigations on the uniform component of the Galactic magnetic field are complicated by the random component of the field, whose strength exceeds that of the uniform component (for instance in Phno & Shibata (1993) the non-random component is estimated to be $\sim 5\ \mu$G with a scale length for fluctuations ~ 100 pc). Various techniques have been applied to determine the magnetic field. E.g. in the detailed analysis based on the pulsar rotation and dispersion measures carried out in Rand & Lyne (1994), a local field strength of 1.4 ± 0.2 μG and direction $\theta_0 = 88^o \pm 5^o$ are found. Some problems arise if one considers a magnetic model of concentric circular field lines as in Rand & Lyne (1994). In fact in other galaxies it has been observed that the galactic magnetic field closely follows the spiral configuration. The work presented in Heiles (1996) develops this spiral-frame line of research which is based on the large-scale data set on starlight polarization (Mathewson & Ford, 1970) with nearly 7000 stars. The advantage of this kind of data is that they are free of systematic errors and the polarization is accompanied by the source location and estimate of extinction. We use the magnetic field parameterization by Strong et al. (2000)

$$B_{tot} = B_0 e^{-(R-R_\odot)/R_B - |z|/z_B}. \tag{3.18}$$

The parameters B_0, R_B and z_B are chosen to best reproduce the Haslam et al. (1982) 408 MHz all-sky continuum survey, which combines data from four different surveys using Jodrell Bank MkI, Bonn 100 meter, Parkes 64 meter and Jodrell Bank MkIA telescopes.

Random fluctuations are not included in the model, although they might contribute to the observed synchrotron emission.

3.3.6 Diffusion Coefficients

In Section 2.4.4 the spatial diffusion coefficient 2.136 and momentum diffusion coefficient 2.137 have been deduced for a charged particle which is scattered by random hydromagnetic waves propagating along a regular magnetic field $\mathcal{H}_0$. Here we assume that the CR plasma obtains a stationary state in a fixed spectrum of turbulences. We restrict ourselves to a power law energy spectrum in wavenumber k so that wave energy density is

$$W(k) = \frac{w\mathcal{H}_0^2 L}{4\pi 91 - a}(kL)^{(-2+a)}, \; kL \leq 1, \; a = const., \tag{3.19}$$

where

$$w = \frac{4\pi}{\mathcal{H}_0^2} \int_{1/L} W(k)dk, \tag{3.20}$$

characterizes the turbulence level which is equal to the ratio of magnetohydrodynamic wave energy density to magnetic field energy density; L is the principal scale of the turbulence.

Substituting 3.19 in 2.136 and 2.137 yields

$$D = \frac{2}{3\pi}\frac{1-a}{a(2+a)}\frac{vL}{w}(\frac{r_g}{L})^a, \tag{3.21}$$

and

$$D_{pp} = p^2\frac{v_A^2}{vL}\frac{2\pi}{(1-a)(2-a)(4-a)}\left(\frac{r_g}{L}\right)^{-a}, \tag{3.22}$$

where $r_g = v/|\omega_H|$ is the particle's gyroradius. The parameter w in Eq. 3.21 characterizes the level of turbulence, i.e. the ratio of magnetohydrodynamic (MHD) wave energy density to magnetic field energy density. Remind that the diffusion coefficient D has been derived for transport along the large scale magnetic field $\mathcal{H}_0$, i.e. only the component $D_{ii} = D_{||}$ is nonzero. It is usually assumed that some large scale wandering of the magnetic field creates a large scale random magnetic field and consequently the other diagonal elements of the diffusion tensor become non-zero. In the limit of isotropic diffusion, i.e. $D_{ii} = D$ for all i, the spatial diffusion coefficient 3.21 is reduced by a factor 3, to account for this large scale wandering of the magnetic field lines. Note, that the limit of isotropic diffusion is a strong simplification. For a realistic scenario one would expect that the propagation along the unperturbed field is still more pronounced than perpendicular to this field. This has also been impressively demonstrated by following the path of a single CR in a perturbed magnetic field for various scenarios (De Marco et al., 2007). Clearly, anisotropic diffusion should dominate CR diffusion in most regions of the Galaxy. An exception may be regions with a very high level turbulence such as SN shells. However, only very few state-of-the-art transport models actually allow for anisotropic diffusion. The reason for this is that the detailed magnetic field structure of our Galaxy is unknown and that anisotropic diffusion adds at least two additional free parameters to the transport model, which, given the fact that all transport parameters have almost no observational constraints at all, leads to more paramters than constraints. For now we will follow the isotropic approach and return to the more realistic anisotropic diffusion in Chapter 4. From Eqs. 3.21 and 3.22 we find

$$D_{pp}D = \frac{4p^2v_A^2}{3a(4-a^2)(4-a)w}, \tag{3.23}$$

which means that that the momentum diffusion coefficient is given by the spatial diffusion coefficient and v_α. Note that the diffusion coefficient has been reduced by a factor 3 compared to Eq. 3.21. Since only v_A^2/w is relevant here, one can subsume $w = 1$ in v_α and treat v_A^2 as an effective parameter. After analyzing all the quantities in Eq. 3.21 one can

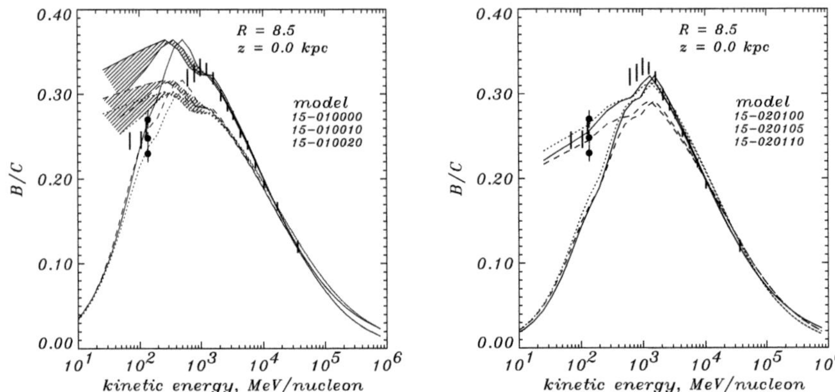

Figure 3.5: B/C ratio for DC models without **(left)** and with **(right)** break in diffusion coefficient, for $dV/dz = 0$ (solid line), 5 (dotted line), and 10 km s^{-1} kpc^{-1} (dashed line). The halo height is in all cases $z_h = 10$ kpc. Solid lines: interstellar ratio, shaded area: modulated to 300 - 500 MV. Data: as in Fig. 3.6. The figures are reproduced from Strong & Moskalenko (1998), the model numbers shown in the figures refer the the GALPROP parameter files used by the authors.

argue that a convenient form for the spatial diffusion coefficient is

$$D = \beta D_0 (\frac{\rho}{\rho_0})^{\delta}, \tag{3.24}$$

where ρ is the rigidity, ρ_0 is a reference rigidity introduced for an eventual break, D_0 is a normalization factor and δ is a free parameter of the model, which can be different below and above the break. Finally, we have reduced the problem of diffusion to two fundamental parameter D_0 and δ (plus ρ_0 in case of a break) while diffusive reacceleration is connected to diffusion via v_α.

3.4 The Isotropic GALPROP Models

Despite the many simplifications like the homogeneous gas distribution or the isotropic diffusion coefficient, the most important observations like the secondary to primary ratios and the local CR spectra can be described in the isotropic GALPROP models. On the

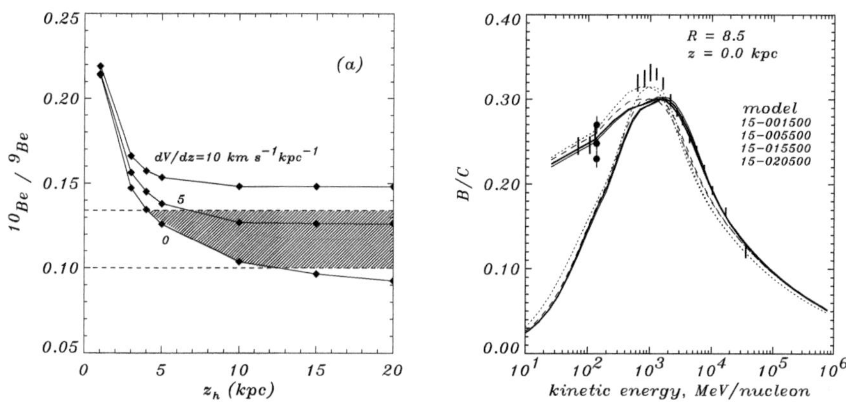

Figure 3.6: Left: Predicted $^{10}Be/^{9}Be$ ratio as function of z_h for $dV/dz = 0,\ 5,\ 10$ kms^{-1}kpc^{-1} at 525 MeV/nucleon corresponding to the mean interstellar value for the Ulysses data Connell (1998); the Ulysses experimental limits are shown as horizontal dashed lines. The shaded regions show the parameter ranges allowed by the data. From Strong & Moskalenko (1998). **Right:** B/C ratio for diffusive reacceleration models: $z_h = 1$ (dotted), 5 (dashed), 10 (thin solid), and 20 kpc (thick solid). In each case the interstellar ratio and the ratio modulated to 500 MV is shown. Data: vertical bars: HEAO-3, Voyager (Webber et al., 1996), filled circles: Ulysses (DuVernois et al., 1996), Φ= 600, 840, 1080 MV.

other hand isotropic CR transport has rigorous limits which we will discuss in Section 3.5. Here we briefly describe the predictions of the isotropic GALPROP models, of which various branches exist.

The propagation parameters are usually tuned to the secondary/primary ratio and the unstable/stable ratio of locally observed charged particles, while the injection spectra are chosen to best reproduce the local proton and electron spectra.

Plain Diffusion (PD) Models

The most simple model one can think of is a plain diffusion model with no diffusive reacceleration and no convection. This model is very similar to a leaky-box approach, except that here we can also take care of the case of slow diffusion, i.e. the steady state CR distribution is not flat in spatial coordinates. The price to pay for this simplicity is an ad-hoc break in the rigidity dependence of the diffusion coefficient and an additional factor of β^{-3} that needs to be introduced to match the B/C ratio at low energies. Specifically, in this model $D = \beta^{-2} D_0(\rho/\rho_0)^\delta$ with $\delta = 0$ below and $\delta = 0.6$ above the reference rigidity $\rho_0 = 3\,\mathrm{GV}$ and $D_0 = 2.2 \cdot 10^{28}\,\mathrm{cm}^2\,\mathrm{s}^{-1}$. A good fit to the B/C and $^{10}Be/^{9}Be$ ratio is obtained for a halo height of $z_h = 4$ kpc.

Diffusion Convection (DC) Models

Here diffusive reacceleration is neglected in favor of convection. The basic parameters are the diffusion coefficient and the convection velocity gradient dV/dz. The convection velocity at $z = 0$ is assumed to be negligible. In such models a good fit to the data is not possible, because the characteristic peaked shape of the measured B/C ratio cannot be reproduced. The left side of Fig. 3.5 shows the B/C ratio for $z_h = 10$ km. In order to force a fit to the data, one can assume an *ad hoc* break in the diffusion coefficient, this is also shown in Fig. 3.5.

From the $^{10}Be/^{9}Be$ ratio the maximum gradient in convection velocity can be limited to 7 km/s for $z_h \geq 4$ kpc. The left side of Fig. 3.6 summarizes the limits on z_h and dV/dz using the $^{10}Be/^{9}Be$ ratio at the kinetic energy of 525 MeV/nucleon appropriate to the Ulysses data (Connell, 1998). Obviously, larger convection velocities require larger haloes, which is clear considering that convection increases the particle flux in Eq. 2.143 and the escape time of CR increases for increasing halo heights.

Diffusion Reacceleration (DR) Models

If diffusive reacceleration (i.e. 2nd order Fermi acceleration as described in Section 2.3.2) is included, a good description of the data can be found for $v_A = 30$ km s^{-1} and no break in diffusion coefficient (Strong & Moskalenko, 1998). The diffusion coefficient has the well motivated form $D = \beta D_0(\rho/\rho_0)^\delta$ with $\delta = 1/3$ and $D_0 = 1.7 - 16 \cdot 10^{28}\,\mathrm{cm}^2\,\mathrm{s}^{-1}$ for $z_h = 1 - 20$ kpc. Reacceleration provides a natural mechanism for reproducing the energy dependence of the B/C ratio as shown on the right side of Fig. 3.6. It should be noted that statements about propagation parameters must always be considered in the context of a given model, for example the Kolmogorov spectrum ($\delta = 1/3$) is ruled out in the semi-analytical model studied in Maurin et al. (2001).

DC models, which are quasi-isotropic, since they incorporate a weak anisotropy via the weak convection velocity, have difficulties in accounting for the observed B/C ratio, unless one assumes an *ad hoc* form of the diffusion coefficient. On the other hand models with reacceleration account naturally for the energy dependence over the whole observed range, with only two free parameters. Combining these results points rather strongly towards the reacceleration picture.

3.5 Limits of Isotropic Transport Models and Evidence for Anisotropic CR Transport

The isotropic transport models already provide reasonable predictions of the local CR fluxes. They are based on a number of priors which can be formulated in the following way:

1. propagation is homogeneous and dominated by isotropic pitch-angle scattering, in particular the scattering rate is assumed to be the same in the halo and in the disk

2. convective transport modes are assumed to be negligible and any R-dependence in convective transport is neglected

3. the gas in the disk is distributed homogeneously, leading to an equal weight for the interaction rate with the different components of the gas.

These priors fulfill the basic picture of the origin and propagation of cosmic rays discussed in Chapter 2.

However, isotropic transport models come with severe caveats. First of all, the assumption of isotropic, spatially constant diffusion, is at odds with what one would expect from the ISM. From a practical point of view the reduction of the spatially dependent diffusion tensor to a single scalar constant is a great advantage, but this procedure is generally considered an oversimplification. Secondly, there are a number of observations which are incompatible with the assumption of isotropic or quasi-isotropic transport, such as the ROSAT observations of Galactic winds, the soft γ-ray gradient as observed by EGRET and COS-B, the large Bulge over Disk ratio in positrons and the exclusion of thermal positrons from the molecular phase of the ISM as observed by INTEGRAL. These observations and the corresponding implications for CR transport will be discussed in the following.

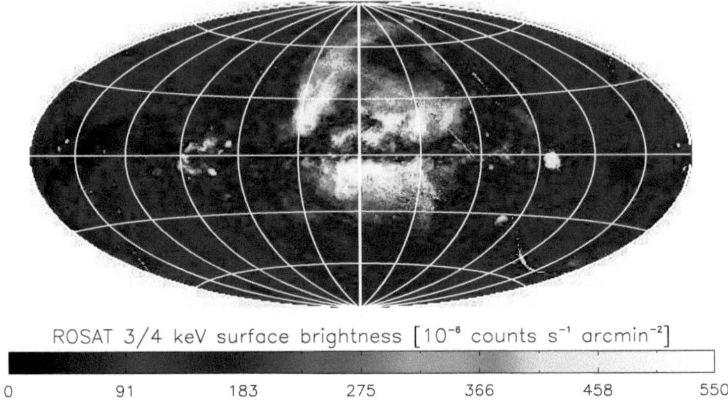

Figure 3.7: X-ray emission at 3/4 keV (the "R45 band") as seen by ROSAT (Snowden et al., 1997). These observations suggest a "Galactic X-ray Bulge", seen most clearly in the southern Galactic Hemisphere, and stretching over the Galactic longitude range, l, from $|l| \leq 30^o$ and down to approximately -15° in Galactic latitude. This X-ray bulge in the southern Galactic Hemisphere can be explained with a combined thermal and cosmic-ray driven wind. From Everett et al. (2008)

3.5.1 The ROSAT Galactic Wind Observations

Isotropic diffusion models are generally incompatible with vertical wind velocity gradients larger than 10 km/s/kpc (Strong et al., 2007), because for larger wind velocities the CRs are blown into the halo and the required ratio of times spent in the halo and the disk from the combined constraints from local secondary to primary ratios and radioactive cosmic clocks cannot be met. Figure 3.6 shows the $^{10}Be/^{9}Be$ ratio for different convection velocities in an isotropic diffusion model. Clearly, a wind speed of around 7 km/s/kpc seems to be the limit for the GALPROP models. Until recently there was no direct observational evidence that the Milky Way's atmosphere might feature winds at all. It has been suggested that the Milky Way's SN rate is too small to build up sufficient CR pressure to overcome the Galactic gravitational potential. In this respect, the fact that current transport models are incompatible with significant convection velocities has not been considered a serious deficiency. This changed in 2007 when it was found that the Milky Way drives a large scale wind too: the ROSAT-satellite observed an enhancement of the diffuse soft X-ray

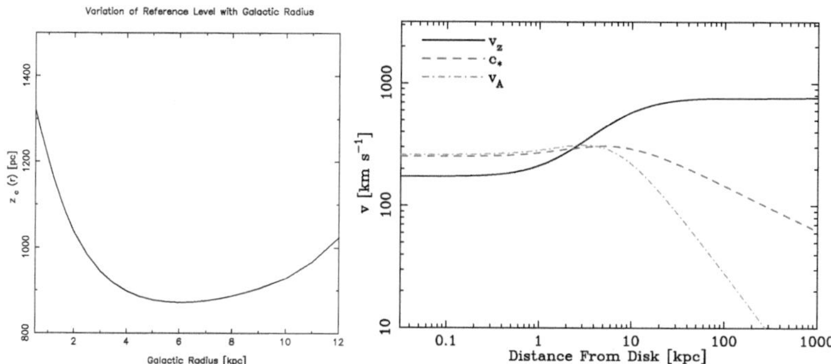

Figure 3.8: Left: Dependence of the diffusion-convection boundary, $z_c(R)$, on Galactocentric radius R. The location of z_c at any radius is defined by the balance of diffusive and convective CR fluxes. The curve shown here has been obtained from self-consistent Galactic wind calculations. From Breitschwerdt et al. (2002). **Right:** Velocity vs. height in a fiducial wind model. The solid line represents the wind velocity, the dashed line represents c_*, the composite sound velocity, and the dot-dashed line shows the change in the Alfvén velocity with height. This velocity curve shows the rather standard increase in velocity of a pressure-driven wind, rising from the relatively low v_0, through the critical point at $v = c_*$, and accelerating on to $v \sim v_\infty$. From Everett et al. (2008).

background emission (Levenson et al., 1997). Figure 3.7 shows the X-ray sky as observed by ROSAT. The emission can be explained best by a mixed CR and thermally-driven wind model (Everett et al., 2008). Their model is based on a model by Breitschwerdt et al. (2002), where the spatial shape of the wind velocity is given by the balance of the gravitational potential and CR pressure and basically follows the SNR distribution. Breitschwerdt et al. (2002) demonstrated that the Milky Way can launch winds, if the dynamical coupling between the escaping cosmic rays and the thermal plasma is taken into account. Since the CR energy density is with 1.5 eV/cm^3 (Webber, 1987) comparable to the energy density of galactic magnetic field and to the energy density of turbulent motions of interstellar gas, CRs are an important dynamical factor in the Galaxy. Breitschwerdt et al. (2002) carried out a self-consistent calculation of Galactic winds. Given a dependence of the cosmic ray source distribution on Galactocentric radius R, the numerical wind solutions show that the wind velocity depends both on R, as well as on vertical distance z. In regions with a high density of CR sources the CR pressure is strong enough to overcome the gravitational forces and drives gas into the halo. The convection velocity increases with increasing height above the plane, which means that for some distance z_c the convective transport will become stronger than the diffusive transport. Following Jokipii (1976) the convection-diffusion boundary z_c can be defined to lowest order by

$$z_c(R) \sim \frac{D_{zz}(R, z_c)}{V_c(R, z_c)} \tag{3.25}$$

Above this boundary convection dominates and the probability for a CR to return to the diffusion region below z_c decreases exponentially. In the presence of CR sources, where it is easy to overcome the gravitational potential, CRs will leave the Galaxy earlier and consequently the interaction rate will be smaller than e.g. in the outer Galaxy. In such a model grammage and escape time depend on both particle rigidity and Galactocentric radius.

The local CR escape time is then given by

$$\tau_{esc} \sim \frac{z_c^2}{D_{zz}} \sim \frac{D_{zz}}{V_c^2}. \tag{3.26}$$

If the wind velocity depends on Galactocentric distance the diffusion-convection boundary will also depend on R, thus significantly changing the shape of our effective diffusion box. The left side of Fig. 3.8 shows the convection-diffusion boundary as calculated by Breitschwerdt et al. (2002). The functional dependence of $z_c(R)$ with radius is straightforward to understand. Close to the Galactic center, the gravitational pull is strongest. Since the authors of Breitschwerdt et al. (2002) chose all other quantities being the same across the disk (constant density, thermal pressure, magnetic field strength), the outflow velocity, and hence mass loss rate, is smallest in the Galactic center. Equation 3.25 then tells us, that $z_c(R)$ must be largest. In the outer parts of the Galaxy the source strength and hence the CR pressure decreases. Consequently the outflow velocity also decreases and $z_c(R)$ must increase again. It is noteworthy that the minimum of $z_c(R)$ at $R \sim 6$ kpc does not coincide with the maximum of the source distribution (see Fig. 3.3) at $R \sim 3-4$ *kpc*. This is probably a consequence of the interplay between the gravitational field and the source distribution in the fully nonlinear equations (cf. Breitschwerdt et al. (1991)). Breitschwerdt et al. (2002) chose a constant thermal pressure throughout the Galaxy. In reality one should also take into account the thermal temperature and pressure in regions of higher supernova activity. The net effect would be a more pronounced peak in outflow velocity and a deeper minimum in $z_c(R)$, respectively, and therefore an even better quantitative

proportionality between $V_c(R)$ and the source distribution. Therefore we find it reasonable to assume that the convection velocity is proportional to the source distribution.
Returning to the ROSAT observations, Everett et al. (2008) found wind velocities from 173 km/s in the disk to 760 km/s in the halo from a fit of their thermally driven wind model as can be seen from the right side of Fig. 3.8. Compared to starburst galaxies which feature wind velocities up to 3000 km/s these wind velocities are still moderate. However, the impact of even moderate convection velocities of a few 100 km/s on CR transport is significant as we will show in Section 4.3.
In the isotropic models which can allow only for very small convection velocities the diffusion-convection boundary is basically defined by the free escape boundary condition at a fixed height z_h above the plane, which leads to a constant CR escape time throughout the disk. The ROSAT observations constitute strong evidence of anisotropic transport by virtue of a significant convective transport mode.

3.5.2 The COS-B and EGRET Soft γ-Ray Gradient Observations

It is interesting to note that the model suggested by Breitschwerdt was not motivated by direct evidence for Galactic winds, but by another problem of isotropic transport models: The distribution of supernova remnants, which are believed to be the sources of CRs, peaks toward the Galactic center, as shown in Fig. 3.3. In an isotropic diffusion model the propagated CR distribution still strongly resembles the source distribution, leading to a peak in the radial distribution of diffuse γ-rays, i.e. one observes a strong gradient, as shown on the left side of Fig. 3.9. This is incompatible with the soft γ-ray gradient as observed by COS-B and EGRET[3]. In order to reproduce the observed γ-ray gradient, a source distribution significantly flatter than the observed distribution of supernova remnants has to be chosen (Strong & Moskalenko, 1998) (see the right side of Fig. 3.9, the flattened source distribution is also shown in Fig. 3.3). Breitschwerdt et al. (2002) found that the discrepancy between the CR source distribution and the diffuse Galactic γ-rays can be explained entirely by propagation effects, if dynamical coupling between the escaping CRs and the thermal plasma is accounted for. From Eq. 3.26 it is clear that the CR escape time decreases with increasing source strength, if the convection velocity is proportional to the source strength. This leads to an R-dependent CR interaction rate and consequently a mild γ-ray gradient.

[3]Note, that this γ-ray gradient refers to a spatial feature of the diffuse γ-rays. A possible miscalibration of the EGRET instrument will not modify the spatial shape of the observed radiation.

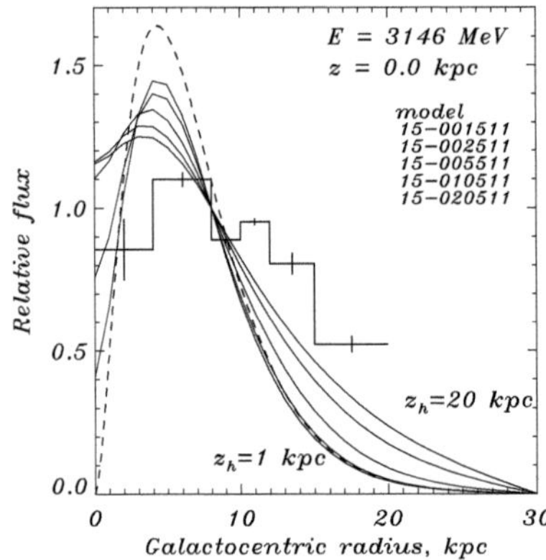

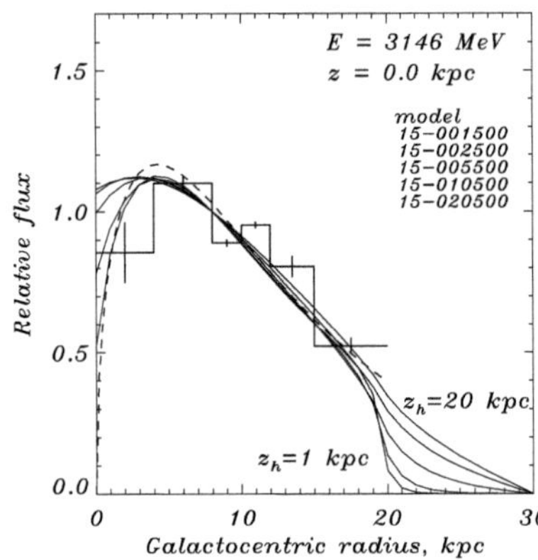

Figure 3.9: **Left:** Radial distribution of 3 GeV protons at $z = 0$, for diffusive reacceleration model with halo sizes $z_h = 1$, 3, 5, 10, 15, and 20 kpc (solid curves). The source distribution is that for SNR given by Case & Bhattacharya (1996), shown as a dashed line. The cosmic-ray distribution deduced from EGRET >100 MeV gamma rays (Strong & Mattox, 1996) is shown as the histogram. **Right:** Radial distribution of 3 GeV protons at $z = 0$, for diffusive reacceleration model with halo sizes $z_h = 1$, 3, 5, 10, 15, and 20 kpc (solid curves). The source distribution used is shown as a dashed line, and is that adopted to reproduce the cosmic-ray distribution deduced from EGRET >100 MeV gamma rays (Strong & Mattox, 1996), shown as the histogram. From Strong & Moskalenko (1998).

As an alternative to anisotropic transport Strong et al. (2004b) investigated the possibility of a strong increase in X_{CO} on the diffuse γ-rays. X_{CO} is the conversion factor from the CO integrated temperature to the H_2 column density. The $\lambda = 2.6\ nm$ ($J = 1 \rightarrow 0$) emission line from carbon monoxide (CO) is a tracer for the spatial distribution of H_2 molecules, because collisions between the CO and H_2 molecules in the clouds are responsible for the excitation of carbon monoxide. This means that any variation in X_{CO} only applies to the molecular component of the gas, not to the atomic HI and HII components. The X_{CO} conversion factor can be determined either by methods based on the assumption of molecular cloud virialization, which suffer from large uncertainties, or from γ-ray analyses. The second of these methods is much more reliable, however, we would like to point out, that this method only holds if one assumes that CRs penetrate molecular clouds freely and

produce secondaries such as γ-rays there. Taking the observed pulsar distribution, which is thought to be a good tracer of the SNR distribution, as CR source function, an isotropic transport model is able to reproduce the observed γ-rays only if a strong increase of the X_{CO} scaling factor with Galactocentric radius R is assumed. The reasoning behind this is, that X_{CO} increases if the metallicity decreases. Since the metallicity dependence of X_{CO} is logarithmic a significant decrease in metallicity is required to change X_{CO}.
Diffuse γ-rays stem from interactions with all components of the ISM as well as from inverse Compton scattering and so the variations in the X_{CO} factor have to be rather large in order to have a significant effect. Diffuse Galactic γ-rays require an increase in X_{CO} of a factor of 25 between 3 kpc and 15 kpc and still a factor of about 12.5 from 6 kpc to 11 kpc. In addition the electron flux has to be scaled by an additional factor of 0.7. This means that the local electrons are a factor 2.8 too high in a model with increasing X_{CO} (compared to a factor 4 for a model with flattened source distribution).
Although an increase of X_{CO} with Galactocentric radius is expected from the observed metallicity gradient, the authors of Strong et al. (2004b) found that the gradient has to be rather large in order to be compatible with the diffuse γ-ray data.
When examining the impact of Galactic winds on CR transport, we assume a constant X_{CO} factor in this work as a first step and allow for an increase with Galactocentric radius only if required.

3.5.3 The Size of the Transport Box

The halo size is one of the most important parameters for an isotropic transport model, since it severely limits the allowed transport parameters. Generally, diffusion coefficients and convection velocities can only be quoted for a given halo height. The reason for this is that the diffusion coefficient and the convection velocity determine the CR flux as given in Eq. 2.143. For a given halo height this flux has to be chosen in such a way, that the constraints from the local secondary to primary ratios and the radioactive instable isotopes are met. Isotropic transport models prefer halo heights slightly above 4 kpc. Figure 3.1 shows the changes in the locally measured B/C and $^{10}Be/^{9}Be$ ratio for an increase in halo height, while the transport parameters are fixed. In GALPROP the halo height is given by the boundary applied to the numerical solution of the transport equation. Beyond this boundary the CR density is put to zero, which basically means that free escape is assumed. Since the CR density in the intergalactic space is non-zero, this is obviously only an approximation, which can be relaxed in models with a physical treatment of the boundary condition. For an isotropic diffusion model with constant diffusion coefficient this means that one assumes the density of scattering centers to suddenly decrease to zero beyond the boundary, while inside the transport box the density of scattering centers is constant. Of course a smooth decrease in the scattering centers would be a much more natural solution, but this requires an increase in diffusion coefficient towards the boundary. This is exactly what is needed in anisotropic propagation models, as will be discussed in Sect. 4.1. In such anisotropic models the boundary of the diffusion box can be in principle at infinity in contrast to isotropic models, where the residence time increases with the size of the diffusion box for a constant diffusion coefficient.

3.5.4 The INTEGRAL 511 keV Line

Recent observations of the positron annihilation line by the INTEGRAL instrument onboard the NASA SPI mission led to surprising insights into transport phenomena in our Galaxy. INTEGRAL observed the positron annihilation signal from the Galactic center (Weidenspointner et al., 2007). The main source of low energetic positrons are SNIa which are distributed in the bulge and in the disk. Strikingly, the annihilation signal in the disk can be entirely explained by the decay of ^{26}Al from core collapsed SNs, leaving no room for additional positrons from SNIa (Prantzos, 2006). Low energy positrons can annihilate with electrons, preferentially bound to nuclei in order to prevent a large momentum transfer during the collision. A detailed account of the annihilation process was given by Guessoum et al. (2005). Such positrons largely originate from the decay of radioactive nuclei expelled by dying stars. In the case of SNIa the SNR core makes up a large fraction of the mass, so it has a relatively thin layer of ejecta, which makes it easier for the positrons to escape. Light curves, which are sustained first by the γ-rays in the shock waves and later by the electrons and positrons, suggest that only a few percent of the positrons escape from the ejecta and can annihilate outside after thermalization. Positrons annihilating inside the ejecta will also produce γ-rays, but these will not be visible as an annihilation line due to the successive interactions in the shock wave.

Positron Large Bulge over Disk Ratio

INTEGRAL found a very strong 511 keV positron annihilation line towards the Galactic center corresponding to an annihilation rate of $(1.5 \pm 0.1) \cdot 10^{43} s^{-1}$ (Knödlseder et al., 2005). In contrast, the annihilation signal from the disk was very weak. A bulge/disk

(B/D) ratio of a few was quoted, although additional data show a clear disk signal as well (Weidenspointner et al., 2007). Taking the dominant source to be β^+ decay of ^{56}Co in SNIa, one would expect a B/D ratio to be well below one, because of the higher mass in the disk and the higher rate of SNIa explosions expected in the thick disk as compared to the bulge (Prantzos, 2006). An additional problem presents the observation of the 1.8 MeV line from the ^{26}Al radioactive isotope, which was clearly observed in the bulge *and* the disk by the Comptel detector on NASA's CGRO observatory (Diehl et al., 2006). These nuclei are thought to be produced by nucleosynthesis in massive stars and yield in their decay on average 0.85 positrons. The observed flux of positron annihilation in the disk seems to be saturated already by the positrons from ^{26}Al and ^{44}Ti, leaving little room for additional positrons from SNIa explosions in the disk. Remember that ^{26}Al has a half life of the order of $7.2 \cdot 10^5$ years, so the positrons from their decay are not convected away by the wind from the dying star. In isotropic transport models this result cannot be understood, because the $\sim$ MeV positrons do not propagate ($D_{zz} = D_{RR} \propto \beta \cdot \rho^{\delta}$), so that positrons annihilate close to their sources in the bulge and in the disk.
In isotropic models the observed large bulge/disk ratio is only explainable if an additional positron population is assumed. This positron population has to be confined to the bulge in order to reproduce the observed B/D ratio and in addition the positrons have to be assumed to annihilate close to their sources. Several possible candidates have been proposed, which are able to either partially (see, e.g. the low-mass-X-ray-binaries in Weidenspointner et al. (2008)) or entirely (see, e.g. the contribution from dark matter annihilation in Ascasibar et al. (2006)) explain the observed signal from the bulge. However, these explanations cannot account for the absence of a positron annihilation signal from the positron from the SNIa population. Prantzos (2006) pointed out that B/D ratios as small as 0.5 are compatible with the INTEGRAL data if the disk positrons diffuse sufficiently away from their sources. He estimated that about 50% of the $\sim$ MeV positrons have to leave the confinement region below z_c before slowing down.
In Subsection 4.3.3 it will be shown that transport models including convection can explain the large B/D ratio in a natural way.

Exclusion of Positrons from Molecular Clouds

Diffusion models generally assume that all transport parameters are global parameters and that the gas is homogeneously distributed. As long as only the γ-rays, which give the

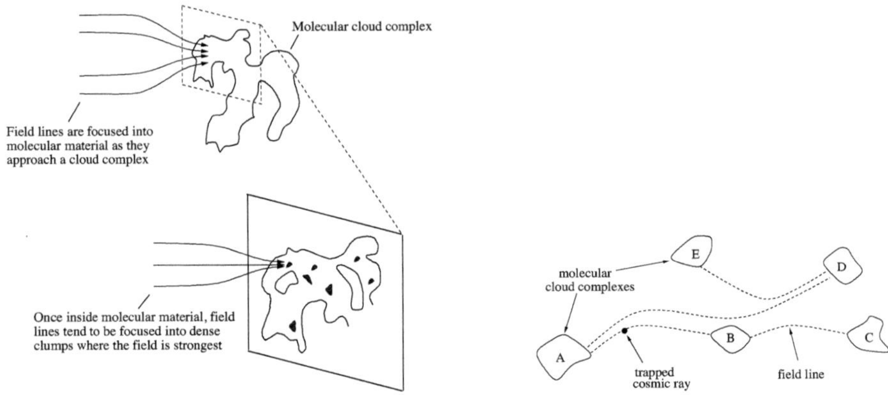

Figure 3.10: Graphical representation of magnetic field line focussing in MCCs and CR trapping. From Chandran (2001)

averaged emission along a line of sight are of interest, this is a reasonable assumption. If the local fluxes of CRs are considered, then local deviations have to be included in a model.

In addition to the large B/D ratio the INTEGRAL instrument measured the spectral morphology of the annihilation line. The positron annihilation predominantly seems to take place in the diffuse non-molecular phase of the ISM. The annihilation signal from the molecular component, which makes up about 40% of the total gas mass of our Galaxy, appears to be compatible with zero (Jean et al., 2006). A possible explanation might be the small volume filling factor of the H_2 clouds of only a few % (Launhardt et al., 2002). The rest of the gas is homogeneously distributed and constitutes the intercloud material. As mentioned in Section 3.1 the isotropic transport models assume that the gas distribution is not clumpy, i.e. the interaction rate in the different components is defined by the average density, leading to a large number of interactions in the molecular component.

The magnetic field in molecular clouds (MCs) is proportional to the cloud's gas density and ranges from 6 μG to 120 μG (Troland & Heiles, 1986) while the magnetic field in the intercloud material (ICM) is $4 - 5\mu G$ (Zweibel & Heiles, 1997). Since magnetic field lines are focussed into the strong field regions a magnetic field line passing through a molecular cloud complex has a higher probability of entering a MC than a straight line of sight (see Fig. 3.10). For molecular cloud complexes (MCCs) this means that there is more magnetic flux through the smaller dense clumps than through the complex as a whole. A CR will be magnetically reflected by a cloud complex if its pitch angle cosine fulfills

$$\frac{v_{||}}{v_{\perp}} < \sqrt{\frac{B_{max}}{B_{ICM}} - 1}, \tag{3.27}$$

where B_{ICM} is the field strength in the ICM and B_{max} is the maximum field strength a CR encounters inside a cloud complex and $v_{||}$ is the velocity component parallel to the magnetic field. Thus molecular clouds, like any high field region, can magnetically reflect CRs and act as magnetic mirrors. If the mean distance of magnetic mirrors is smaller or of the order of the mean diffusive scattering length for a CR, MCs can act as magnetic traps and confine CRs to the volume between MCs until their pitch-angle no longer fulfills equation 3.27. A well-understood example of such a process are the so called van-Allen-belts of our Earths magnetic field.

If MCCs are indeed efficient at trapping CRs, then this would mean that CRs can spend time in the intercloud material and possibly undergo radioactive decays or interact with the intercloud material without changing their spatial distribution due to diffusive scattering. Such a process cannot be modelled in a pure diffusion model, because it occurs on scales below or comparable to the mean scattering length. It can be incorporated by scaling the energy losses and gains, as well as the secondary interaction rate accordingly. However, such a scaling is complicated, because the details of this scaling factor and especially its energy dependence are unknown. We show the impact of magnetic traps on a diffusion model in Appendix D in more detail, but due to the unknown details we will neglect trapping by MCCs in the following.

Other Inhomogeneities of the ISM

There are other inhomogeneities in the ISM that are likely to affect CR transport, especially the secondary to primary ratios. Examples are the Local Bubble, a low density region surrounding the Sun, and the spiral structure of the Milky Way. It is not known in how far these structures have an impact on CR transport and none of the contemporary transport models is able to embed these structures in the calculations. Here we are interested in the consequences of Galactic winds mainly and therefore we will neglect these features for now and in the next chapter we will show that it is possible to be compatible with wind speeds as expected from ROSAT. In this course we will develop a program which is actually capable to model structures like the Local Bubble and the spiral arms and we will address these issues in Chapter 5.

Chapter 4

An Anisotropic Transport Model for Galactic Cosmic Rays

In the past years new observations as well as theoretical considerations have modified and complicated our picture of CR transport significantly, as discussed in Section 3.5 in detail before:

- Self-consistent Galactic wind calculations have led to the conclusion that convective transport plays a non-negligible role for CR transport. In such models the size of the diffusion dominated zone in the Milky Way changes with Galactocentric distance (Breitschwerdt et al., 2002). Recently, an analysis of the ROSAT X-ray observations (Everett et al., 2008) has confirmed this picture, as discussed by Breitschwerdt (2008).
- The large scale distribution of the positron annihilation signal as observed by INTEGRAL shows a large bulge to disk (B/D) ratio, which is strong evidence for a propagation effect, because even the addition of an -unknown- positron source cannot explain why there is almost no annihilation signal in the disk.

Isotropic diffusion models for Galactic cosmic ray transport put tight constraints on the maximum convection velocity in the halo. For a half halo height of 4 kpc the maximum convection speed is limited to 40 km/s in the halo, since otherwise the constraints from local secondary to primary ratios and radioactive isotopes cannot be met. The ROSAT Galactic wind observations of wind speeds up to 760 km/s therefore constitute a problem for diffusion models.
Here it is shown that such wind speeds are possible, if the diffusion coefficient in the halo

is different from the diffusion coefficient in the disk. The radial dependence of the wind velocity was taken to be proportional to the source strength, as expected from winds which are sustained by cosmic ray pressure. In this case the cosmic ray density and with it the diffuse γ-ray production from nuclear interactions are suppressed near the sources. This solves in a natural way the problem of the soft gradient in the radial dependence of the γ-ray flux. Furthermore, the absence of an annihilation signal from disk positrons from SNIa as observed by INTEGRAL, can be explained by positron escape from the disk in such a model.

In this chapter we develop an anisotropic model for CR transport (anisotropic propagation model (aPM)) which is capable of explaining both local and interstellar CRs with a common source spectrum. The model meets the observational constraints from ROSAT and INTEGRAL. The collection volume for CRs in such a model is comparable to the collection volume of the isotropic models and thus the local grammage and CR age is correctly estimated. The model was realized by adapting the GALPROP code, which is now the most detailed and most powerful tool for CR transport that we have.

This chapter is organized as follows: In Section 4.1 we will describe the minimal extensions of the isotropic transport model which are necessary to account for the observations described above. Section 4.2 describes the parameter determination for this new model and sections 4.3 and 4.4 present the model's predictions in terms of local charged CRs and γ-rays, respectively. We will end this chapter with a few remarks on diffuse γ-rays.

4.1 Minimal Modifications of the Transport Equation

From the discussion in Section 3.5 we find that any realistic transport model should at least allow for the following features:

1. Wind velocities as expected from the self-consistent Galactic wind calculations, the soft γ-ray gradient (Breitschwerdt et al., 2002) and the ROSAT data (Breitschwerdt, 2008; Everett et al., 2008)

2. spatial inhomogeneities in all transport parameters

3. possibly anisotropic diffusion

4. spatial inhomogeneities in the ISM.

To this end the GALPROP code was modified in the following way:

- In GALPROP an equidistant grid is used to numerically solve the diffusion equation. However, for parameters varying on small scales a fine grid is required, which would dramatically increase the memory requirements and computing time. Therefore a non-equidistant user-defined spatial grid with arbitrary grid points was implemented to allow for a course grid spacing in the halo with simultaneously a fine grid in the disk.
- The isotropic diffusion coefficient is replaced by D_{RR} for transport in the R direction and D_{zz} for transport in the z direction. Both diffusion coefficients may depend on spatial coordinates and may have an independent energy dependence.
- The convection velocity may depend on Galactocentric radius.
- To model local gradients in transport parameters and gas density the user can specify regions (possibly corresponding to the Local Bubble and the Local Fluff (Frisch, 2009)) for which the transport parameters or gas density may differ from the global parameters.
- The spatial derivatives of the propagation parameters, which are needed to solve the diffusion equation, have been calculated for the additional R and z dependence of the diffusion and convection parameters. The corresponding Crank-Nicholson coefficients are given in the Appendices B.1 and B.2.

The parameterization of the spatial dependence of the convection and diffusion will be discussed in the following sections.

4.1.1 Convection Velocity

Convection was chosen to be proportional to the R-dependence of the source density to take care of the increased CR pressure close to the maximum of the source distribution. The convection velocity is parameterized as follows:

$$V_c(R,z) = Q(R,0)(\Theta(|z| - z_0 \cdot V_0 + \frac{dV}{dz}z). \tag{4.1}$$

Here $Q(R,0)$ is the R-dependence of the CR source distribution, z_0 is the initial height from where the wind is launched, Θ is the Heaviside step function, V_0 is the convection velocity at $z = z_0$ and dV/dz the gradient of the convection velocity. A linear increase in z was chosen for reasons of simplicity. The source distribution was adopted from Case

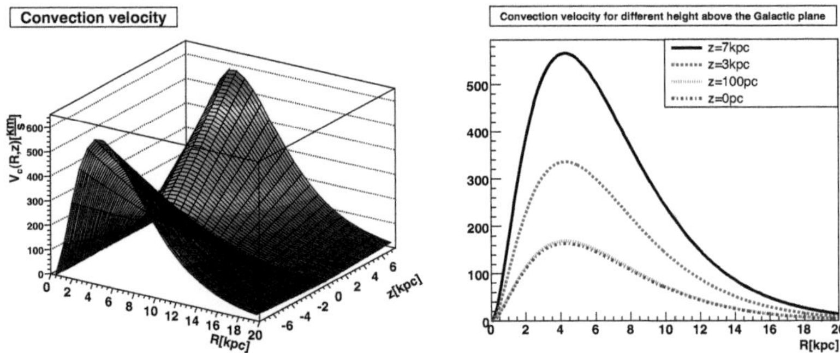

Figure 4.1: **Left** Convection velocity in an aPM.$V_c(R, z) = Q(R, 0)(V_0 + z(dV/dz))$, with $V_0 =$ 100 km/s and $dV/dz = 35$ km/s and $Q(R, 0)$ given by the SNR distribution by Case & Bhattacharya (1996) (normalized to 1 at R_o). **Right** Convection velocity for different distances from the Galactic plane. Below z=0.1 kpc only the contribution from zdV/dz plays a role. The maximum wind velocity was chosen to be 591 km/s (in good agreement with ROSAT).

& Bhattacharya (1996), the parametrisation is given in Eq. 3.15 where we adopted their values $\alpha = 1.69$, $\beta = 3.22$ and $z_s = 0.2$ kpc.
The ROSAT data indicates an initial velocity of 173k m/s at the base of the wind, see Fig. 3.8. Therefore $z_0 = 0$ kpc with a local start velocity of 100 km/s was chosen, which results in a start velocity of about 170 km/s at the peak of the SNR distribution. dV/dz was chosen to be 35 km/s/kpc leading to a maximum convection velocity of 591 km/s at $z_h = 7.5$ kpc and 733 km/s at 16 kpc, which is in good agreement with ROSAT. The spatial distribution of the convection velocity for these parameters is shown in Fig. 4.1.
Note, that Everett et al. (2008) found that for Galactocentric radii smaller than 1.5 kpc no wind can be launched due to the high gravitational potential. They used a simple rectangular function to model the R-dependence of the Galactic wind. In this work we assume that the strong decrease in source distribution and thus convection velocity towards the Galactic center is sufficient to model the influence of the Galactic gravitational potential.

4.1.2 Diffusion Coefficients

The spatial dependence of the diffusion tensor is not well constrained. There have been efforts to calculate the detailed behavior of the diffusion coefficient in the halo (see. e.g. Dogiel et al. (1994); Dogiel & Gurevich (1993)), but generally one would expect an increase of the diffusion coefficient in the halo from the observed isotropy of CRs, the amount of local ^{10}Be and the vertical CR distribution (see e.g. Ginzburg et al. (1980); Ptuskin & Khazan (1976)). Here, a linear vertical increase in diffusion coefficients is chosen. This choice correctly reproduces the slope of the secondary to primary ratios as we will see in Section 4.3. Any possible R-dependence is neglected.

In the disk diffusion is assumed to be constant. In this model we use a rather large zone of constant diffusion with a half height of $z_d = 1$ kpc. The diffusion coefficients are parameterized as

$$D_{zz} = D_{RR} = \begin{cases} \beta D_0 \left(\frac{\rho}{\rho_0}\right)^{\delta}, & |z| < z_d, \\ \beta D_0 \frac{|z|}{z_d} \left(\frac{\rho}{\rho_0}\right)^{\delta}, & |z| \geq z_d, \end{cases} \tag{4.2}$$

where ρ_0 is the reference rigidity, D_0 is a proportionality constant, which is treated as a free parameter, δ is the slope of the power law describing the rigidity dependence of the diffusion coefficient and $\beta = v/c$ is the particle velocity. The parameter δ is taken to be the same for all rigidities, i.e. there is no break in the power law of the rigidity dependence. In the following we assume that the regular magnetic field plays no role, although this is clearly a questionable assumption. In principle one would expect an anisotropy in diffusion coefficients, because even in the presence of strong scattering, the large scale component of the Galactic magnetic field is non-negligible (Codino & Plouin, 2007; De Marco et al., 2007). In the model presented here we choose $D_{zz} = D_{RR}$, which implies locally an isotropic scattering, but the overall scattering rate decreases towards the boundary by the positive gradient in D_{zz} and D_{RR}. However, a more detailed model for CR transport should also take care of the structure of the unperturbed magnetic field.

4.1.3 Diffusion-Convection Boundary

For a large enough gradient in convection, convective transport will dominate over diffusion above a certain value in z. In Section 3.5.1 the convection-diffusion boundary z_c was defined as (Eq. 3.25) $z_c(R) \sim D_{zz}(R, z_c)/V_c(R, z_c)$. Above this boundary convection dominates and the probability for a CR to return to the diffusion region below z_c decreases exponentially. The R and z-dependence of the parameters are explicitly included. For the parameters of the anisotropic propagation model as determined in Section 4.3, the R-dependence of z_c is plotted in Fig. 4.2 for different rigidities. For most radii the regions of confinement where diffusion dominates, i.e. below the curves, are much smaller than the halo boundary, set

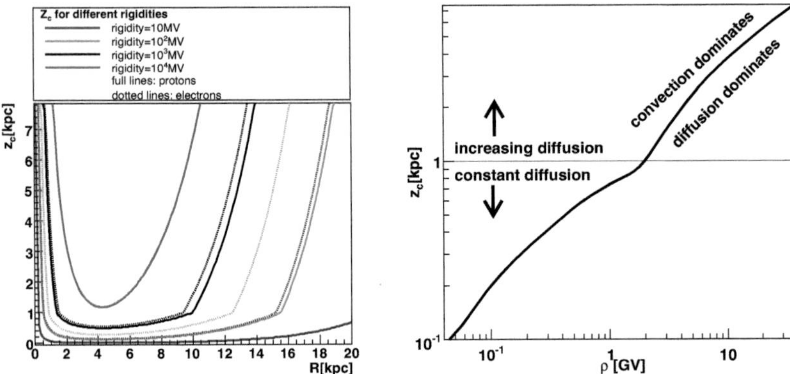

Figure 4.2: Left Convection diffusion boundary as defined by $D_{zz}(R, z_c) = V_z(R, z_c) \cdot z_c$ in an aPM for different particle rigidities. The *full lines* correspond to protons, the *dotted lines* correspond to electrons. For rigidities larger than $10^4 MV$ D_{zz} and therefore z_c is the same for protons and electrons. The kink at 1kpc above the plane is caused by the sudden increase of the diffusion coefficient in this model (see Subsection 4.1 for details). **Right** The rigidity dependence of the convection-diffusion boundary z_c for $R = 8.3$ kpc as defined by $D_{zz}(8.3 \text{ kpc}, z_c, \rho) = V_c(8.3 \text{ kpc}, z_c) \cdot z_c$. The change in slope at 1 kpc corresponds to the transition from constant diffusion to z-dependent diffusion.

to 7.5 kpc. Note that near the GC the convection becomes small by virtue of the decrease in the source distribution. Such a strong decrease is also expected from first principles, because of the strong gravitational potential in the Galactic Center (GC), which will inhibit the launch of Galactic winds. Figure 4.2 also shows the rigidity dependence of the diffusion-convection boundary for protons. The increase in diffusion for distances larger than 1 kpc leads to a change in slope of $z_c\rho$ around 1 GeV. In Section 4.2 we will see that this change in slope is the key point to reproducing the high energy slope of the local B/C ratio in the presence of Galactic winds.

Parameter	aPM	Conventional Model
Injection Spectra		
Protons/ nuclei		
$\alpha_1/\alpha_2/\alpha_3$[1]	1.6/1.8/2.41	1.98/2.42
ρ_1^p/ρ_2^p	2 GV/9 GV	9 GV
Electrons		
β_1/β_2	1.6/2.54	1.6/2.54
ρ_1^e	4 GV	4 GV
Injection spectra are of the form $\left(\frac{\rho}{\rho_i}\right)^{-(\alpha_i,\beta_i)}$		
Transport Parameters		
D_0	$5.3 \cdot 10^{28}\ \frac{\mathrm{cm}^2}{\mathrm{s}}$	$5.8 \cdot 10^{28}\ \frac{\mathrm{cm}^2}{s}$
ρ_0	4 GV	4 GV
δ	0.33	0.33
z_d	1 kpc	-
V_0	$100\ \frac{\mathrm{km}}{\mathrm{s}}$	-
z_0	0 kpc	-
$\frac{dV}{dz}$	$35\ \frac{km}{s \cdot kpc}$	-
$V_c(R)$	$\propto Q(R)$	-
v_α	$56\frac{km}{s}$	$30\frac{km}{s}$
B_0	6.5 μG	6.1 μG

Table 4.1: Parameters of the aPM and a conventional GALPROP model.

4.2 Parameter Determination for the Anisotropic Propagation Model (aPM)

In this section we discuss the parameter tuning for the aPM proposed in Section 4.1. It is not the aim of this study to present a fine-tuned best fit model, which anyway would be rather short-lived in the light of the upcoming Fermi-LAT and PAMELA data releases, but rather to show that the ROSAT Galactic winds in principle are compatible with local CR measurements.

The optimization of the parameters follows the same path as for the isotropic model discussed in Chapter 3, i.e. the diffusion coefficients are chosen to best reproduce the local

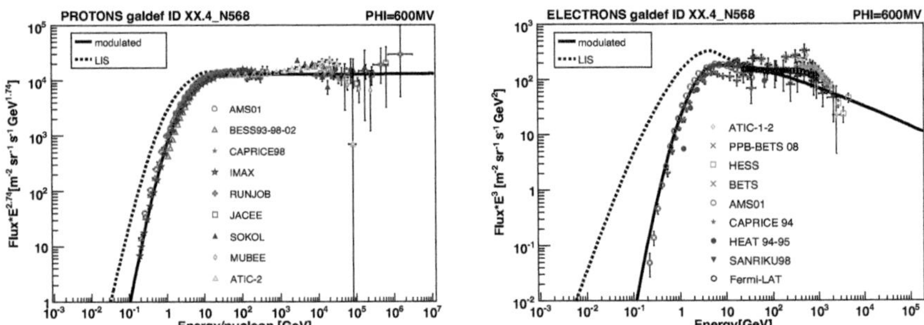

Figure 4.3: Left Proton flux in an aPM: local proton flux (*full line*) and LIS proton flux (*dashed line*). **Right** Electron flux in a conventional aPM. Line coding as on the left. Data are taken from the CR database by Strong & Moskalenko (2009).

B/C and $^{10}Be/^{9}Be$ ratio and the injection spectra for protons and electrons are chosen to fit the local proton and electron spectra. The convection velocity parameters are taken from the ROSAT data, as discussed above in Section 4.1.1. The most important transport parameters for this model are summarized in Table 4.1 and will be discussed in more detail below.

The left side of Fig. 4.3 shows the local protons spectrum (at Earth) and the local interstellar proton (LIS) spectrum. The correction for solar modulation was done in the force-field approximation (Gleeson & Axford, 1968), as implemented in the GALPLOT program [2].

The right side of Fig. 4.3 shows the local electron spectrum. The electron injection index was chosen to give a propagated electron spectrum with an index of roughly 3.3. This yields a somewhat softer spectrum than observed by Fermi (Abdo et al., 2009), but agrees with an extrapolation of the low energy electron data. The reasoning behind this is that for high energies local sources may contribute significantly to the local electron flux, because the large electron synchrotron losses cool electrons from distant sources efficiently. These sources are not included in our calculations, which means that one *expects* our propagated electron spectrum to be somewhat softer than the data indicates. We will discuss possible local sources that might contribute in this energy range in Chapter 5.

In order to cope with the large CR transport time and the comparably small amount of secondaries, CRs have to spend a certain time in the halo. The times spent in the halo and the disk, denoted by t_h and t_d, respectively, are constrained by the ratio of

[2]The GALPLOT routine is available from http://www.mpe.mpg.de/~aws/propagate.html.

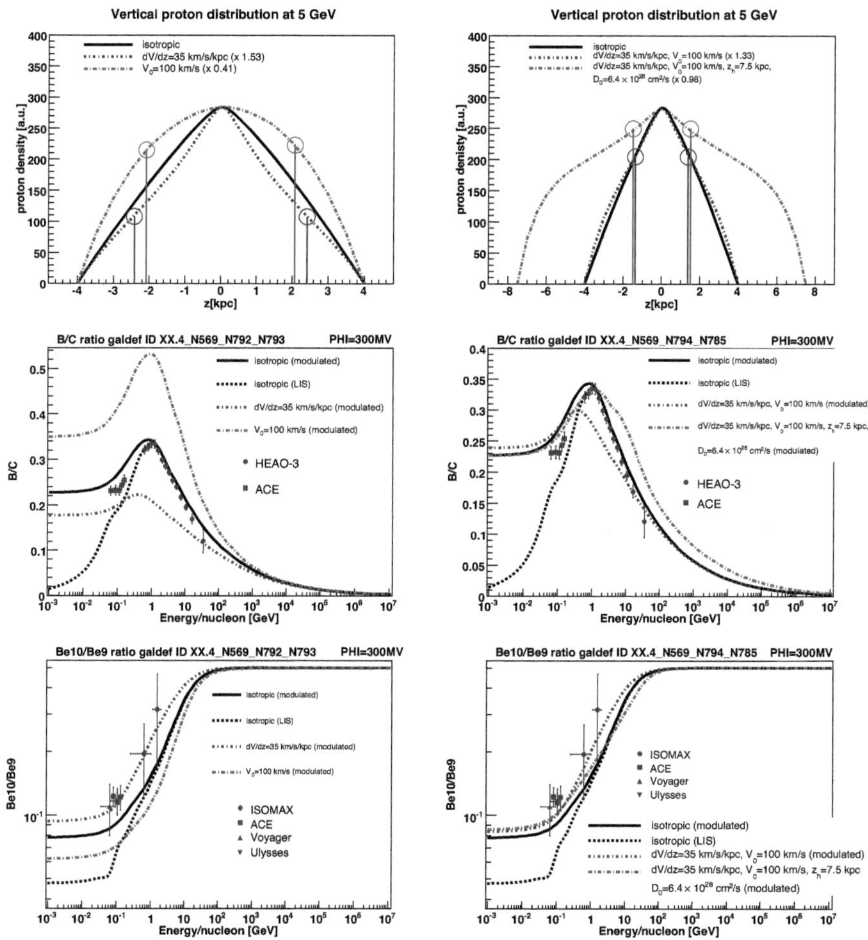

Figure 4.4: **Left column:** The influence of dV/dz and V_0 on the vertical proton distribution in arbitrary units (**top**), B/C (**center**) and $^{10}Be/^{9}Be$ (**bottom**). The vertical proton distribution has been normalized to the local density of the isotropic run to allow for better comparison, the normalization factors are given in parenthesis. The open circles indicate the position of the diffusion-convection boundary z_c for the different models. For the isotropic model z_c is identical to the halo boundary at 4 kpc. *Data for* B/C*:* HEAO-3 (Engelmann et al., 1990) ACE (Davis, 2000). *Data for* $^{10}Be/^{9}Be$*:* ISOMAX (Hams et al., 2004), ACE (Yanasak et al., 2001), Voyager (Lukasiak, 1999), Ulysses (Connell et al., 1998). **Right column:** Same as the left column for a model with Galactic winds as expected from ROSAT (blue short dashed-dotted line) and a model with Galactic winds and increased halo height and diffusion coefficient. Data for B/C and $^{10}Be/^{9}Be$ as in the left column.

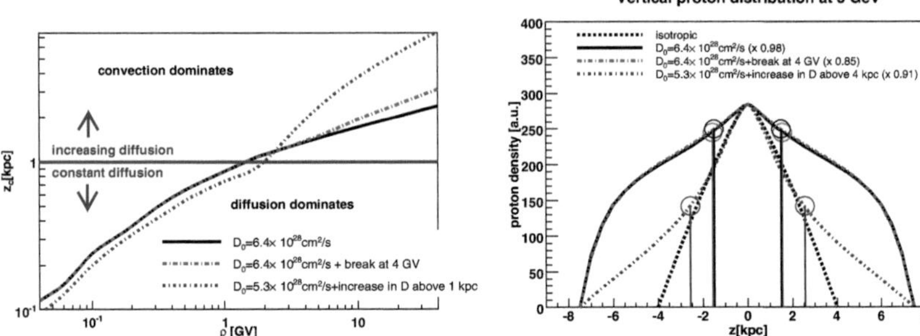

Figure 4.5: Left: Rigidity dependence of the diffusion-convection boundary z_c for a model with constant diffusion coefficient (black full line), a model with broken diffusion coefficient (red long dashed-dotted line) and a model with increasing diffusion coefficient in the halo (blue short dashed-dotted line). For the model with break in diffusion coefficient we use $\delta = 0.33$ below 4 GV and $\delta = 0.5$ above 4 GV. D_0 is kept at $6.4 \cdot 10^{28}$ cm^2/s and the proton injection index is with 2.25 above 9 GV somewhat harder than the conventional model with 2.42 to counteract the additional momentum losses above 4 GV. For the model with increasing diffusion coefficient we use $D_{zz} = D_{RR} = 5.3 \cdot 10^{28}\mathrm{cm}^2/\mathrm{s} \cdot (\rho/4\ \mathrm{GV})^{0.33}$ for $z \leq 1$ kpc and $D_{zz} = D_{RR} = 5.3 \cdot 10^{28}\mathrm{cm}^2/\mathrm{s} \cdot (|z|/\mathrm{kpc}) \cdot (\rho/4\ \mathrm{GV})^{0.33}$ for $|z| > 1$ kpc. The change in slope at 1 and 4 kpc corresponds to the transition from constant diffusion to z-dependent diffusion (indicated by the blue full line) and the transition from $\delta = 0.33$ to 0.5, respectively. **Right:** Vertical proton distribution at 5 GeV for the three models compared to an isotropic model. In order to allow for easy comparison, the proton density has been normalized to the isotropic model, the normalization factor are given in parenthesis.

secondary/primary CRs, which is most precisely measured for the B/C ratio and the ratio of unstable/stable CRs, which is most precisely measured for the $^{10}Be/^{9}Be$ ratio. Isotropic transport models can reproduce the observed B/C and $^{10}Be/^{9}Be$ ratios, because the diffusion coefficient and halo size can always be chosen in such a way that $t_d/(t_h + t_d)$ agrees with the measurements.

To understand how it is possible to invoke convection velocities as expected from ROSAT we examine the impact of convection on the vertical proton distribution, shown in the left column of Fig. 4.4. Starting from a pure diffusion model (black full line) an increase in the convection velocity gradient dV/dz from 0 to 35 km/s/kpc narrows the proton distribution (blue short dashed-dotted line). This leads to a reduction of grammage and a corresponding decrease of CR age, as can be seen from the B/C and $^{10}Be/^{9}Be$ ratio in the left column

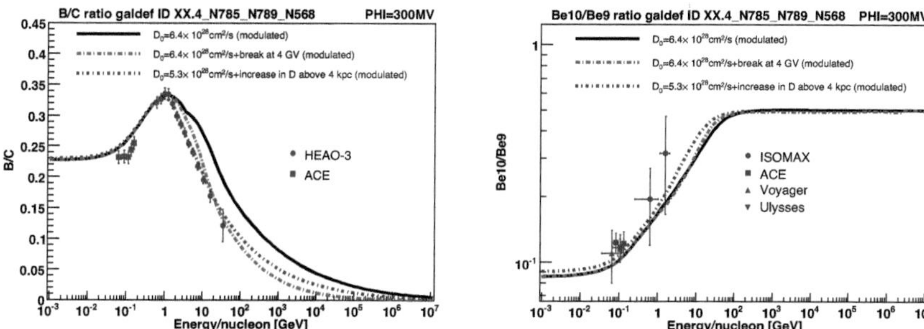

Figure 4.6: *B/C* ratio (**left**) and $^{10}Be/^{9}Be$ ratio (**right**) for a model with constant diffusion coefficient (black full line), a model with broken diffusion coefficient (red long dashed-dotted line) and a model with increasing diffusion coefficient in the halo (blue short dashed-dotted line). Data as in Fig. 4.6.

of Fig. 4.4. On the other hand, an increase in the initial wind velocity V_0 from 0 to 100 km/s has the opposite effect: the proton distribution is widened and grammage and CR transport time are increased (red long dashed-dotted line). The different behavior of these two parameters becomes understandable if one looks at the corresponding terms in the transport equation. With $V_c = V_0 + dV/dz \cdot z$ and Eq.3.5, CR transport along z can be written as

$$\left[\frac{\partial D_{zz}}{\partial z} - V_o - \frac{dV}{dz}z\right]\frac{\partial \Psi}{\partial z} + D_{zz}\frac{\partial^2 \Psi}{\partial z^2} - \frac{dV}{dz}\Psi = \left(\frac{d\Psi}{dt}\right), \tag{4.3}$$

where we neglected momentum and particle losses and gains. Unlike the vertical gradient dV/dz the initial wind velocity V_0 only occurs in the first term. An *increase* in V_0 can be interpreted as a *decrease* in $\partial D_{zz}/\partial z$, which just corresponds to the reduction of the forward mean scattering length in a moving reference frame. In the examples discussed above $dD_{zz}/dz = 0$, so an increase in V_0 indeed acts as a reduction of the diffusion coefficient with increasing distance from the plane and therefore widens the proton distribution.
Thus, the vertical gradient dV/dz and the initial wind velocity V_0 have opposite effects on the density in the halo. Now we combine them and parameterize convection as described in section 4.1.1 with $V_0 = 100$ km/s and $dV/dz = 35$ m/s/kpc. The resulting vertical proton distribution, B/C and $^{10}Be/^{9}Be$ ratios for this case are shown as the blue short dashed-dotted line in the right column of Fig. 4.4. At 5 GeV the proton distribution resembles the one of the isotropic model. For this energy diffusion dominates up to distances of 1.5 kpc from the plane, as indicated by the vertical lines with the circles on top. However, the

B/C ratio is somewhat too low, which can be remedied by a larger transport box, which leads to a lower vertical gradient in the CR density and a correspondingly higher secondary density in the disc, as shown by the red long dashed-dotted line in the right column of Fig. 4.4). Above 1 GV the ratio is too high, but the $^{10}Be/^{9}Be$ is still acceptable for the chosen diffusion coefficient. To reduce only the high energy part of the B/C distribution one can only play with diffusion, since convection is energy independent. The crucial quantity to look at in this case is the diffusion convection boundary z_c, shown as the black line on the left side of Fig. 4.5. A shift of the diffusion convection boundary towards larger vertical distances above 1 GV (corresponding to an increase in diffusion coefficient) would give rise to faster CR escape for these energies. There are two obvious options to accomplish this: on the one hand a break in the diffusion coefficient leads to a larger diffusion coefficient for higher energies (see the red long dashed-dotted line in Fig. 4.5) and on the other hand an increase of the diffusion coefficient in the halo leads to faster CR escape for higher energies (see the blue short dashed-dotted line in Fig. 4.5). The latter option describes the data on B/C and $^{10}Be/^{9}Be$ somewhat better, as shown in Fig. 4.6, so this option is used in the following with the diffusion coefficient as given in Table 4.1. The increase of the diffusion coefficient in the halo allows one to reduce the diffusion coefficient in the disk by about 20% while keeping the halo height fixed at $z_h = 7.5$ kpc. This shifts the diffusion convection boundary towards the Galactic plane for rigidities smaller than 1 GV (see the left side of Fig. 4.5). Because of the smaller diffusion coefficient in the disk, the model with increasing diffusion coefficient in the halo leads to a very similar vertical proton distribution as in the isotropic model for $z \lesssim 3$ kpc, i.e. below z_c (see right side of Fig. 4.5). Below this boundary CR transport is dominated by diffusion and the different diffusion coefficients ($D_{iso} = 6.2 \cdot 10^{28}$ cm^2/s at 5 GV for the isotropic model and $D < D_{iso}$ below 1.1 kpc, while $D > D_{iso}$ above 1.1 kpc for the anisotropic model) lead to a similar vertical proton distribution. Note, that despite the use of an *isotropic* diffusion coefficient, CR transport in this model itself is *anisotropic* due to the non-negligible convection velocities, hence the choice of the name aPM.

With the basic transport parameters fixed by the local measurements of B/C and $^{10}Be/^{9}Be$ and the CR injection spectra given by the local proton and electron spectra the model parameters are basically settled. Fine-tuning of the transport parameters can be done by considering additional secondary particles, like antiprotons and positrons, which originate from nucleon-nucleon collisions in the ISM.

The left side of Fig. 4.7 shows the local antiproton and positron flux. Antiprotons show

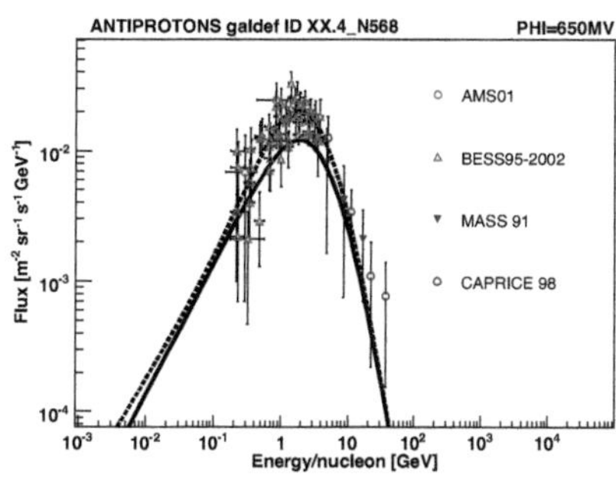

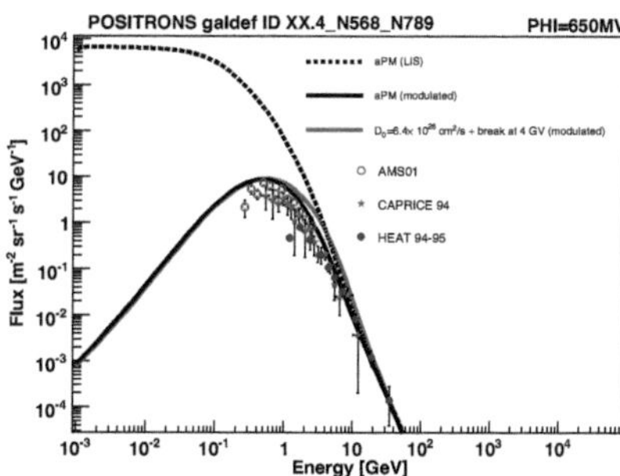

Figure 4.7: Left The antiproton flux in the aPM. The solid line denotes the local flux and the dashed line the flux corrected for solar modulation. *Data:* BESS 95-97 (Orito et al., 2000), CAPRICE 98 (WiZard/CAPRICE Collaboration, 2001), MASS91 (Basini, 1999). **Right** The positron flux in the aPM (black lines). The solid line denotes the local flux and the dashed line the flux corrected for solar modulation. Also shown is the modulated positron flux in a model with break in diffusion coefficient (red line). *Data:* AMS I (Alcaraz et al., 2000), CAPRICE 94 (Boezio et al., 2000), HEAT 94 (DuVernois et al., 2001)

an excess of about 40% to 50% below 4 GeV. This excess is very similar to the excess seen in the isotropic transport models (Strong et al., 2007). It is possible to increase the local antiproton flux by increasing the local CR interaction rate via diffusion or convection. However, this would worsen the B/C ratio and the local positron flux.

The right side of Fig. 4.7 shows the local positron flux. For energies below 5 GeV the model shows a slight excess in local positrons. This is probably the result of too efficient diffusive reacceleration driven by either a too large v_α or a too small diffusion coefficient. On the other hand less efficient diffusive reacceleration would worsen the local B/C ratio. An improvement for both positrons and the local B/C ratio is expected if convective transport was slightly more efficient.

4.3 Performance of the aPM

4.3.1 Halo Size

As discussed before, the diffusion coefficient increases with increasing z, thus providing a natural transition to free escape independent of the boundary condition. For example, increasing the boundary box from z=7.5 to 30 kpc does not change the residence time or

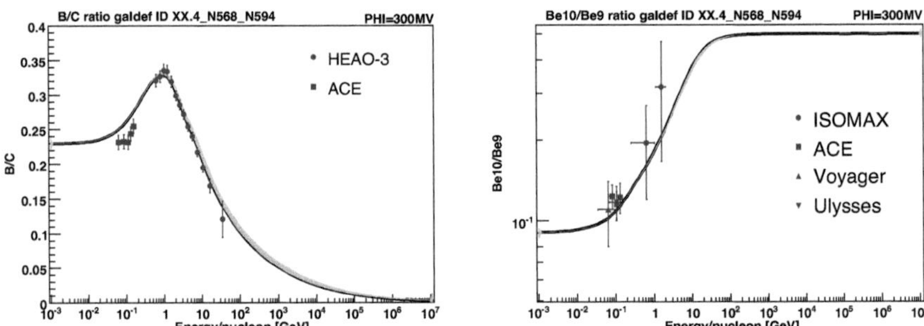

Figure 4.8: Left Local B/C ratio in an aPM (*full line*) and LIS B/C ratio (*dashed line*). *Data:* HEAO-3 (Engelmann et al., 1990) ACE (Davis, 2000). **Right** Local beryllium fraction in a conventional aPM. Line coding as on the left. *Data:* ISOMAX (Hams et al., 2004), ACE (Yanasak et al., 2001), Voyager (Lukasiak, 1999), Ulysses (Connell et al., 1998). The narrow gray band on top of the black line refers to a model with an increased halo height z_h=30 kpc (see Section 4.3.1 for a discussion).

the ratio $t_d/(t_h + t_d)$, as can be seen from the gray band on top of the black line for the B/C ratio in Fig. 4.8, which is hardly distinguishable from the black line for a boundary of z=7.5 kpc. This is in strong contrast to isotropic transport models, which are highly sensitive to the size of the diffusion region, as discussed in Section 3.1 (see Fig. 3.1). Given this large difference in sensitivity to the boundary condition between the isotropic and anisotropic models it is interesting to compare the CR density profiles for protons in the halo. This is done in Fig. 4.9 for two halo sizes in the aPM (z_h = 7.5 and 10 kpc) and two halo sizes in an isotropic model (z_h = 4 and 5.3 kpc) at rigidities of 0.01 GeV and 10 GeV at the Sun's Galactocentric radius. Most sources are located in the disk, so the source density at z = 0 kpc is highest. In the isotropic model low energy CRs are almost at rest and the CRs stay for a large fraction inside the gaseous disk (indicated by the vertical gray band in Fig. 4.9; a disk height of 1 kpc is chosen for illustrative purposes, note that this is not identical to the significantly smaller scaleheight of 250 pc used in the calculation) with tails towards the halo boundary, where the density drops to zero. For the aPM this distribution is broadened at larger z-values by convection. For high energies the situation changes, because the diffusion starts to become more important and the diffusion-convection boundary z_c moves to z-values of a few kpc, as indicated by the hatched area on the right-hand side of Fig. 4.9. The ratio of times spent in the disk and in the halo is approximately given by the area below the curves in the disk and the halo. One observes that this changes for the isotropic model significantly, if one moves the boundary from 4 to 5.3 kpc, while for the aPM the changes are mainly in the convection zone from which only few CRs return to the disk. Together with the increase in diffusion coefficient, convection therefore allows us to reproduce the vertical distribution of the CR distribution in such a way that the requirements from local B/C and $^{10}Be/^{9}Be$ are met and at the same time reduces the model's sensitivity to the exact position of the boundary condition, provided that the diffusion-convection boundary z_c is small enough. Clearly, with increasing energy the convection-diffusion boundary also moves further into the halo until it becomes comparable to the halo height z_h. With increasing energy any aPM will therefore become more and more sensitive to the position of the boundary condition. At the same time however, CRs become less and less confined and for very high energies the diffusion approximation will break down. Figure 4.8 shows that for convection speeds as expected from ROSAT a halo height of 30 kpc still does not yield any significant change on the local B/C and $^{10}Be/^{9}Be$ ratio for all energies of interest, i.e. the energy range in which the diffusion approximation holds.

The radial difference between the two models is less significant, as shown by the two-dimensional proton distribution at 5 GeV in the Rz plane in Fig. 4.10. Both distributions are similar up to Galactocentric radii of 4 kpc. For larger radii the aPM falls off less steep than the isotropic model. The reason for this is that diffusion is assumed to be independent of radius, while convection decreases at larger radii, so diffusion becomes more important there, thus widening the distribution in R. This helps in solving the soft γ-ray gradient problem, as will be discussed in more detail in Section 4.4.2.

4.3.2 Collection Distance

Galactic winds lead to preferred transport in z-direction. As a consequence one could expect, that in the absence of Galactic winds CRs are collected from much larger distances, because they can travel further into the halo and return to the disk. We determine the collection distance of CRs, meaning the maximum distance of CR sources which still contribute to the CR flux measured at a certain point. The collection distance depends on all (local) transport parameters, as well as the local shape of the secondary and primary source distributions.

To examine the R-dependence of the collection distance we consider only sources at a distance d. Note, that this study refers to a two-dimensional run, meaning that the area from where sources can contribute is a ringlike structure with a thickness of $2 \cdot d$ and a

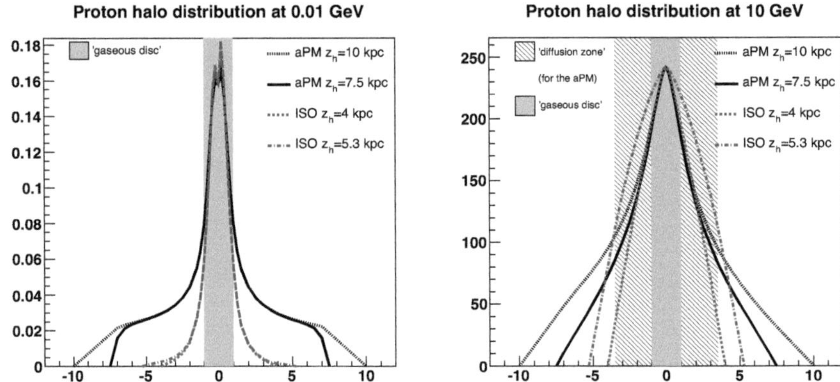

Figure 4.9: The halo density profile of protons for rigidities of 0.01 GeV and 10 Ge at R_o for an aPM (black lines) and an isotropic model (red lines). In order to allow for easy comparison, the proton density for each energy bin has been normalized to the aPM with $z_h = 7.5$ kpc in this figure.

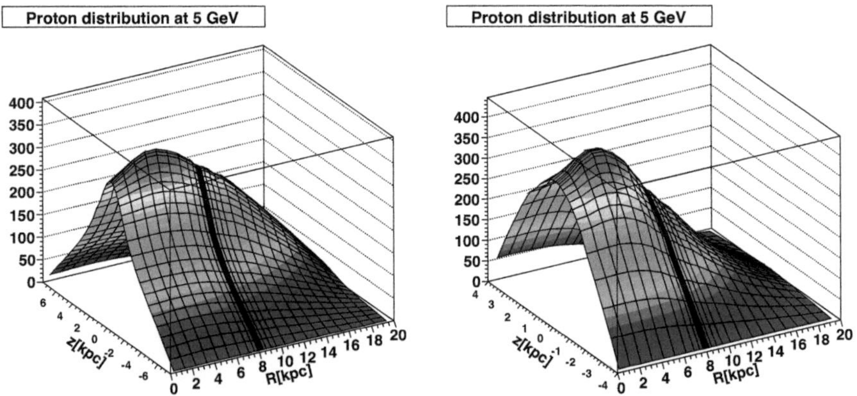

Figure 4.10: The density distribution of 5 GeV protons in the Rz plane for an aPM (left) and an isotropic model (right). Note the different boundaries in z of 7.5 and 4 kpc, respectively. The boundary in R is 20 kpc in both models.

height of $2 \cdot z_H$ =15 kpc. For primaries we apply the cuts directly in the source distribution, for secondaries we consider cuts in the secondary source function. The top left panel of Fig. 4.11 shows the fraction of proton, electron, positron and antiproton flux below 1 GeV at the position of the Earth originating from distances between 0.1 kpc and 12 kpc for an aPM. At these energies the protons are basically at rest (from 1 GeV to 10^{-3} GeV the diffusion coefficient for protons drops by a factor of more than 200) and more than 60 % of the local proton flux stems from distances within 0.1 kpc, while sources within 0.4 kpc contribute with about 90% to the local proton flux. The contribution from sources with distances larger than 2 kpc is insignificant. This local proton population does not produce antiprotons. Due to baryon number conservation the threshold for antiproton production is at about 7 GeV. At threshold the antiproton is produced at rest in the center of mass frame and has an energy of E_{kin} = 938 MeV in the laboratory frame. Antiprotons with $E_{kin} < 938$ MeV have to be produced in the backward hemisphere of higher energy collisions. Most of the antiprotons below 1 GeV are therefore particles which have been produced at higher energies and cooled down on their way to Earth, thus explaining the large collection distance.
When protons become relativistic their collection distance increases significantly, since the diffusion coefficient in proportional to β. Only 10% of the relativistic protons originate from distances within 0.1 kpc and one has to include sources up to a distance of 4 kpc in order to gather 90% of the local protons. For energies larger than 10 GeV the collection distance for protons and antiprotons is comparable (see figure 4.11).
Electrons at 1 GeV are already relativistic, leading to a larger collection distance below 1 GeV. Between 1 GeV and 10 GeV electrons and protons have approximately the same collection distance. Due to their larger energy losses, especially their synchrotron losses, the collection distance for electrons and positrons drops significantly for energies above 10 GeV. Between 10^3 and 10^5 GeV already 70% of the local electrons originate from distances within 0.1 kpc and sources withing 0.2 kpc already contribute more than 90% of the local electron flux.

Figure 4.12 shows the collection distance in the plane for an isotropic model with $D = 5.8 \cdot 10^{28}$ cm^2/s, no convection and a half-halo height of 4 kpc. For all energies, even for the lowest energies, the collection distance is only slightly larger than for the aPM. The reason for this is on the one hand that the diffusion coefficients for both models are still comparable and on the other hand the smaller transport box leads to faster escape in z-direction, thus mimicking the effect of convection. Also shown in figures 4.11 and 4.12 is the uncertainty in collection distance which originates from a change in the local diffusion coefficient.

The collection distance in an aPM is therefore absolutely comparable to an isotropic model. Minor deviations are due to the slightly different diffusion coefficients ($D_{iso} = 5.8 \cdot 10^{28}$ cm^2/s and $D_{aPM} = 5.3 \cdot 10^{28}$ cm^2/s), the large convection velocity does not reduce the radial collection distance of CRs. This means that CRs produced in the peak of the source distribution at 4 kpc, where convection is strongest, will not be immediately blown into the halo, but will still reach Earth with the same probability as in the isotropic model. As the consequence the isotopic composition and the diffuse γ-ray predictions of both models will be similar or only subject to slight variations. We will check the latter in Section 4.4.

4.3.3 The INTEGRAL Positron Annihilation Signal

As discussed in Section 3.5.4 the INTEGRAL satellite observed a large B/D ratio for the positron 511 keV annihilation line. In an aPM this is expected, because low energy particles propagate predominantly by convection, which is large in the disk and small in the bulge. Positrons produced in the disk are transported into the halo, but stay near the sources in the bulge. Furthermore, the bulge has a large extension in all directions, so even if the positrons are slowly transported by diffusion, they still have time to thermalize and find an electron to annihilate. In the disk positrons are produced predominantly in the region of a high source density, i.e. a region of high convection. To quantify the escape probability from the disk the SNR distribution given by equation 3.15 is taken as the source distribution for MeV positrons, but a somewhat larger scale height of 300 pc is used, which is the scale height of SN1a (Prantzos, 2006). These are presumably the supernovae with the highest escape fraction, because of the thinner ejecta. The positron spectrum from ^{56}Co β^+-decays is modeled as a rectangular function between 0.1 MeV and 5 MeV. Figure 4.13 shows the vertical positron distribution between 10^{-6} MeV and 1 MeV in an isotropic model, a model with $V_0 = 100$ km/s, a model with $V_0 = 100$ km/s and $dV/dz = 35$ km/s/kpc and the aPM. The positron density has been normalized to the area below the curve, which means the same source luminosity is assumed in all four cases. Compared to the isotropic model (black full line) the positron density in the disk is reduced by a factor of 2 when a constant wind velocity of $V_0 = 100$ km/s is assumed (blue dotted line). A vertical increase in wind velocity further decreases and slightly widens the positron distribution (blue dash-dotted line). The linearly increasing diffusion coefficient in halo in the case of the aPM (red

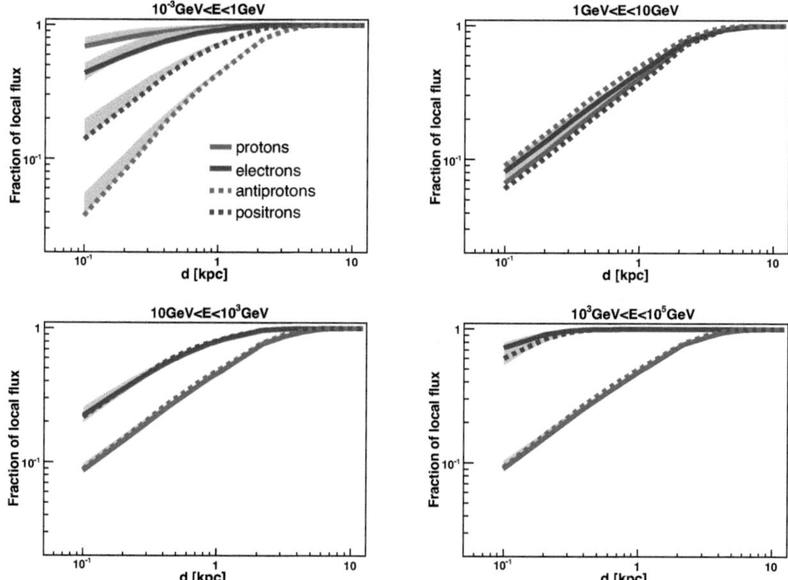

Figure 4.11: Collection distance for protons (full red line), electrons (full blue line), antiprotons (dotted red line) and positrons (dotted blue line) for an aPM for different energy ranges between 10^{-3} GeV to 10^{5} GeV. The grey uncertainty band indicates the change in collection distance due to local variations in the diffusion coefficient in a region with diameter of 0.6 kpc and a height of 0.2 kpc and a diffusion coefficient decreased or increased by a factor of 2. For energies between 10^{-3} GeV and 1 GeV the uncertainties due to a local gradient in diffusion coefficient are shown for primaries and secondaries, for the other energy ranges the uncertainties are shown for primaries only.

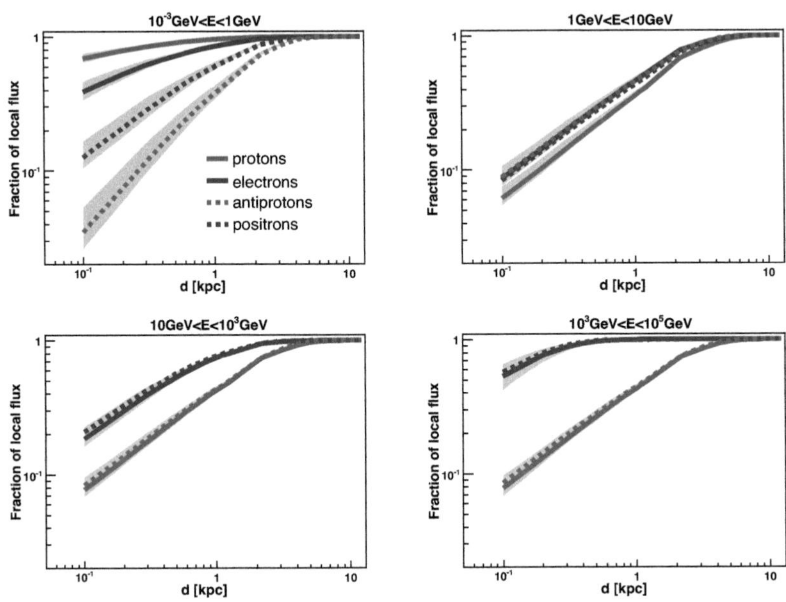

Figure 4.12: Same as fig. 4.11 for an isotropic model with z_h = 4kpc, $D = 5.8 \cdot 10^{28}$ cm^2/s, $v_\alpha = 30$ km/s and no convection.

dashed line) increases the amount of positrons in the halo above 1 kpc (where diffusion starts to increase) and thus leads to a further increase in the positron escape fraction. To further quantify the positron escape fraction in the presence of convection we compare the amount of positrons above and below the diffusion-convection boundary z_c. Table 4.2 shows the fraction of positrons above z_c for different parameters of the aPM. For the aPM more than 89% (83.2%) of the 0.1 (4) MeV positrons escape from the Galaxy. Since CR transport in this energy range is mainly governed by convection, variations of 10% in the diffusion coefficient or a constant diffusion coefficient in the halo (aPM1-aPM3) do not change these numbers significantly. In the disk V_0 is larger than dV/dz, so that a decrease in dV/dz by 50% (aPM4) still yields about the same positron escape fraction. When decreasing V_0 from 100 km/s to 30 km/s for a constant dV/dz = 0 km/s/kpc (aPM5-aPM7) the positron escape fraction decreases significantly and becomes incompatible with the INTEGRAL requirements for V_0 = 10 km/s (ISO1). A strong increase in convection

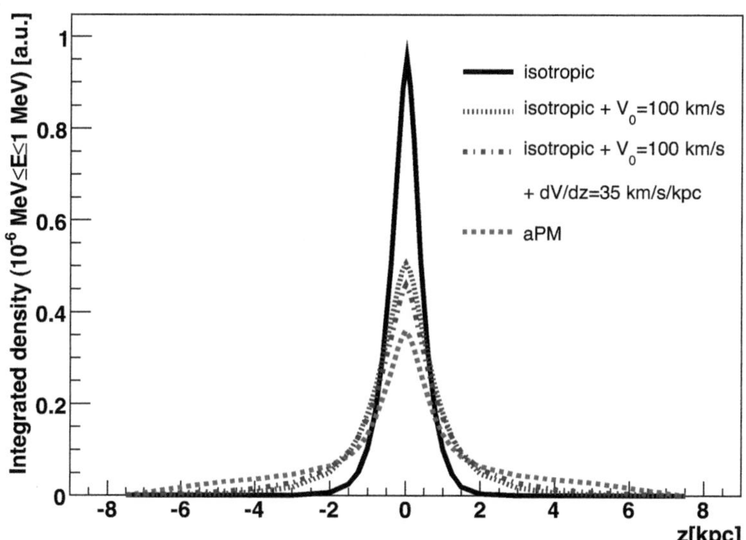

Figure 4.13: The vertical distribution of positrons from SNIa between 10^{-6} MeV and 1 MeV in an isotropic model, a model with $V_0 = 100$ km/s, a model with $V_0 = 100$ km/s and $dV/dz = 35$ km/s/kpc and the aPM. The positron density has been normalized to the area below the curve, which means the same source luminosity is assumed in all four cases

of $dV/dz = 100$ km/s/kpc is compatible with the requirements by INTEGRAL even if the velocity at the base of the wind is zero (aPM8). However in this case the requirements by ROSAT and local charged CRs are no longer met. For a quasi isotropic model with convection velocity of $dV/dz = 7$ km/s/kpc and $z_h = 4$ kpc (ISO) the positron escape fraction ranges from 10.7% at 0.1 MeV to only 3.1% at 4 MeV. Models with V_0 smaller than 30 km/s are generally incompatible with the INTEGRAL requirements.

In summary, models that meet the requirements by ROSAT automatically provide a positron escape fraction high enough to explain the large bulge/disk ratio observed by INTEGRAL. Intermediate convection velocities too small for ROSAT are already able to meet the requirements by INTEGRAL. In the keV range the scale height of the propagated positron distribution is $\sim 1 - 1.5$ kpc, in excellent agreement with the scale height of 1 kpc adopted by Prantzos (2006) in his models B and D. Note, that the positron escape fraction shown here is just a simple estimate. A more detailed modeling of the positron source spectrum would have a significant impact upon the energy dependence of the escape fraction. Furthermore, one has to keep in mind that the diffusion coefficient obtained from a fit to the B/C and $^{10}Be/^{9}Be$ ratio in the GeV range has been extrapolated to MeV energies. This procedure implies that the assumption that CRs scatter on Alvén waves holds down to MeV energies, which is not necessarily true. However, our diffusion coefficient is in rough agreement with the bulge diffusion coefficient derived from contraints on the ISM phase positrons annihilate in Jean et al. (2006). Recently Jean et al. (2009) found that the interaction of MeV positrons with magnetohydrodynamic waves in the neutral phases of the ISM is negligible, which means that positrons move along magnetic field lines with helical trajectories which are perturbed by collisions with particles of the ISM and consequently their diffusion coefficient would be larger than assumed here. On the other hand resonant interaction with Alfvén waves generated by the streaming CRs themselves in the warm interstellar medium might confine positrons for long times to their source regions and thus lead to small diffusion coefficients. In this simple estimate we showed that for diffusion coefficients of the order of 10^{27} cm$^2/s$ [3] an additional convective transport mode can help to explain the observed large B/D ratio.

We have shown that the large positron escape from the disk can be understood if positron transport is taken into account, but it should be noted that due to the smaller source strength in the bulge the number of positrons produced in the bulge is not sufficient to entirely explain the signal from the bulge (Prantzos, 2006). Even in an aPM SNIa may not be enough and an additional process or an additional source population in the bulge may be required to explain the observed emission from the bulge. Prantzos (2006) suggested that a fraction of the positrons escaping the disk may even be channeled by the poloidal field to the bulge where the positrons then would be confined by the large magnetic field. A detailed study of the efficiency of such a channeling process would require to adapt the anisotropy in the diffusion coefficient according to the direction of the magnetic field in the halo and take care of the positron confinement in the bulge by a decreased diffusion coefficient. In this rather qualitative approach we refrain from any fine-tuning of the diffusion coefficients.

Recently, it has been found that an additional low mass x-ray binary (LMXB) population seems to reside in the bulge region showing even the morphological features of the observed

[3]Since $D = \beta D_0(\frac{\rho}{\rho_0})^\delta$ our $D_0 = 5.3 \cdot 10^{28}$ cm$^2/s$ at 4GV leads to an order of magnitude reduction in the MeV range.

annihilation signal (Weidenspointner et al., 2008). As a possible candidate positrons from annihilating (Boehm et al., 2004) or decaying (Hooper & Wang, 2004) light Dark Matter ($m < 100$MeV) have been suggested. However, any additional source in the bulge cannot explain why there is almost no annihilation signal from positrons from ^{56}Co from the disk, but added to an aPM an additional source in the bulge can nicely explain the observe B/D ratio.

With wind velocities taken to be simply proportional to the source distribution we already find excellent agreement with the model presented by Prantzos (2006). The problem of the large B/D ratio for positron annihilation is thus intimately related to the propagation of positrons.

Model	$D_{zz} = D_{RR}$[a] $[cm^2/s]$	dD/dz[b] $[cm^2/s/kpc]$	V_0 $[km/s]$	dV/dz $[km/s/kpc]$	z_h $[kpc]$	f_{esc}@0.1MeV [%]	f_{esc}@1MeV [%]	f_{esc}@4MeV [%]	INT	ROS
aPM3	$1.9 \cdot 10^{27}$	0	100	30	7.5	88.5	85.3	82.1	+	+
aPM1	$1.7 \cdot 10^{27}$	$1.7 \cdot 10^{27}$	100	30	7.5	89.3	87.3	83.3	+	+
aPM	$1.9 \cdot 10^{27}$	$1.9 \cdot 10^{27}$	100	30	7.5	89.2	86.3	83.2	+	+
aPM2	$2.1 \cdot 10^{27}$	$2.1 \cdot 10^{27}$	100	30	7.5	88.0	83.7	81.7	+	+
aPM4	$1.9 \cdot 10^{27}$	$1.9 \cdot 10^{27}$	100	15	7.5	89.0	86.5	82.5	+	+
aPM5	$1.9 \cdot 10^{27}$	$1.9 \cdot 10^{27}$	100	0	7.5	89.0	85.2	82.0	+	-
aPM6	$1.9 \cdot 10^{27}$	$1.9 \cdot 10^{27}$	50	0	7.5	78.0	66.9	65.3	+	-
aPM7	$1.9 \cdot 10^{27}$	$1.9 \cdot 10^{27}$	30	0	7.5	65.9	54.4	47.7	+	-
aPM8	$1.9 \cdot 10^{27}$	$1.9 \cdot 10^{27}$	0	100	7.5	58.6	58.3	56.4	+	-
ISO1	$1.9 \cdot 10^{27}$	$1.9 \cdot 10^{27}$	10	0	7.5	29.1	8.4	0	-	-
ISO2	$1.9 \cdot 10^{27}$	$1.9 \cdot 10^{27}$	0	30	7.5	38.9	34.4	32.0	-	-
ISO3	$1.9 \cdot 10^{27}$	$1.9 \cdot 10^{27}$	0	15	7.5	24.9	18.0	13.4	-	-
ISO4	$1.9 \cdot 10^{27}$	$1.9 \cdot 10^{27}$	0	7	7.5	11.7	4.1	1.7	-	-
ISO	$1.9 \cdot 10^{27}$	0	0	7	4	10.7	4.9	3.1	-	-

Table 4.2: Positron escape fraction for different convection velocities and diffusion coefficients. For details see text.

[a] For $\rho = 1$ MV, $z = 0$
[b] For $\rho = 1$ MV

4.4 γ-rays and Radio Emission

In GALPROP all spectra are internally normalized to units of $\frac{c}{4\pi}$. At the end of the processing of all nuclei the proton flux $\frac{c}{4\pi}n(p)$ at the user-defined reference E_{kin} at $R = R_0$ and $z = 0$ is computed by interpolation to get the correct nomalizing factor as specified in the user defined input file. Since the normalization of all nuclei and the secondary electrons and positrons is linked by the relative source abundances as well as the fragmentation, spallation and decay cross sections the same normalizing factor is applied for all nuclei, positrons and secondary electrons. Primary electrons are normalized separately to a user-defined flux at a user defined energy.

To check the validity of such a self-consistent propagation model at scales outside the kpc scale of the collection volume of charged particles the fluxes of diffuse γ-rays and synchrotron radiation are calculated in GALPROP using the emissivity of the complete diffusion box. The absorption of both γ-rays in the GeV range and synchrotron radiation above a few hundred MHz is small, so that information on the CR density and gas density even in the Galactic Center (GC) can be obtained in the form of line-of-sight integrals, i.e. column densities. The diffuse γ-rays from π^0-decay, bremsstrahlung, IC scattering and synchrotron radiation are calculated self-consistently from the steady-state solution. For this the propagated proton, electron and positron distribution and the respective gas densities, ISRF and magnetic fields, which have already been applied to the calculation of the energy losses are used. The skymaps are produced by integrating the emissivities over the line-of-sight for an observer at the Solar position, for each (l, b) direction in the map. For IC this is straightforward since the volume emissivity is directly calculated using the ISRF and electron, positron spectra. For bremsstrahlung and π^0-decay the emissivity is per nucleon of gas. To best implement the observed Galactic structure in the gas, HI and CO radio-astronomical surveys in Galacticentric rings are used together with the user defined H_2-to-CO (X_{CO}) relation. These data only give column densities per ring, so that the variation of emissivity and gas density within each ring has to be taken into account by some approximation. This is done using a gas-density model as a function of (R, z) as follows:

$$I_\gamma = \sum_i \frac{N_{HI,i} + 2X_{CO,i}W_{CO,i}}{\int\limits_{ring\ i} (n_{HI} + 2n_{H_2})ds} \times \int\limits_{ring\ i} q_\gamma(n_{HI} + 2n_{H_2})ds \qquad (4.4)$$

Region	Longitude l	Latitude $\|b\|$	Description
A	330-30	0-5	Inner Galaxy
B	30-330	0-5	Disk without inner Galaxy
C	90-270	0-10	Outer Galaxy
D	0-360	10-20	Low longitude
E	0-360	20-60	High longitude
F	0-360	60-90	Galactic Poles

Table 4.3: The longitude and latitude of the six sky regions shown in Fig. 4.14.

where i indexes the Galactocentric rings. Here $(n_{HI} + 2n_{H_2})$ is the modelled gas density at any point (R, z) as described in Section 3.3.1, s is the line-of-sight distance and $(N_{HI,i} + 2X_{CO,i}W_{CO,i})$ is the survey-based column density in ring i. The integral is thus corrected for the observed column density while maintaining the model-based variation within the ring. The integrals are performed with a resolution in the line-of-sight distance s of 10 pc.

The calculation of the γ-ray and synchrotron skymaps is done *after* the normalization of the nuclei, electrons and positrons to the local fluxes. Consequently, the γ-ray and synchrotron predictions are also normalized to the local electron and proton density. This is the correct procedure, if the local CR densities are indeed representative for the averaged local proton and electron density, i.e. the CR distribution is sufficiently smooth. If, however, the locally measured proton and electron fluxes represent a local over- or underdensity, then local CRs and the averaged CR densities as observed in diffuse γ-rays and synchrotron radiation have separate normalizations. This can be invoked by just *ad hoc* applying a different normalization as has been done by Strong et al. (2004a) or, more elaborate, by modeling the phenomena which lead to the local deviations, such as variations in the ISM or the source strength. We will discuss this second case in Section 5.3.

4.4.1 Diffuse γ-rays

To our knowledge, there is currently no isotropic transport model in GALPROP able to explain the diffuse γ-rays and the local CRs in a consistent picture. Models are either tuned for local CRs, thus leading to an excess in γ-rays above 1 GeV (*conventional models*) or tuned for γ-rays leading to an incompatibility with local charged CRs by CR density

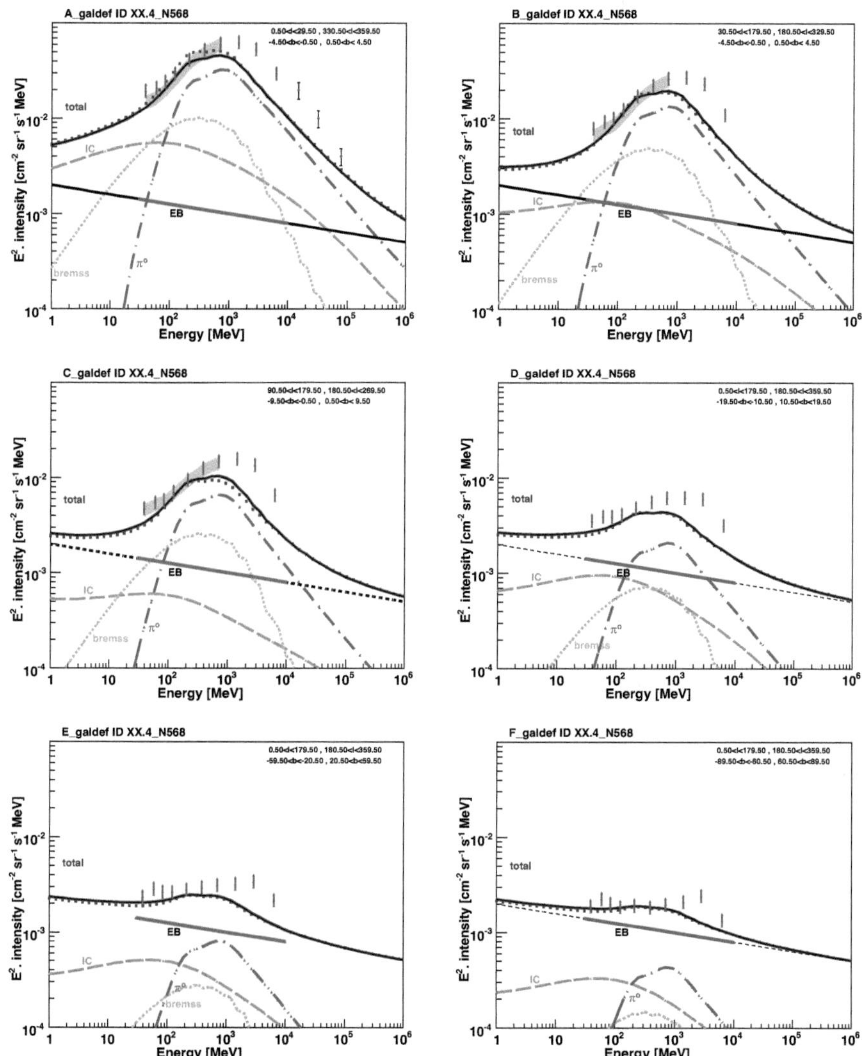

Figure 4.14: Diffuse γ-rays for the six different sky regions as defined in Strong et al. (2004a). Line coding: bremsstrahlung (*light blue dashed*), inverse Compton (*green long dashed*), π^0-decay (*red long dashed-dotted*), total (*blue full*). The pink full line is the extragalactic background model according to Sreekumar et al. (1998). The EGRET data is corrected for the point-spread function (PSF), for regions A, B and C the uncorrected EGRET data is shown as the grey band. Also shown is the total γ-ray flux as predicted by an isotropic model with $D_0 = 5.8 \cdot 10^{28}\mathrm{cm}^2/\mathrm{s}$ at 4 GeV and $v_\alpha = 30\mathrm{km/s}$ (dotted blue line).

and spectral shape (*optimized models*).
In the light of the upcoming release of the Fermi-LAT data on diffuse γ-rays a cross-check with diffuse γ-rays is difficult. The EGRET data does not agree with the preliminary Fermi-LAT data even in the lowest energy range (Porter, 2009). In the absence of other options we continue to use the EGRET data as a cross-check, keeping in mind that the softer Fermi data will probably require a larger contribution from inverse Compton (IC) and bremsstrahlung.

The diffuse Galactic γ-rays are shown in Fig. 4.14, the regions used in this figure are the regions as introduced in Strong et al. (2005). The latitude and longitude ranges are given in Table 4.3. The γ-ray prediction of the aPM in general is similar to the prediction of an isotropic model with the same source distribution and constant X_{CO} (shown in Fig. 4.14 as the dotted blue line). For region A (inner Galaxy) the flux in the isotropic model is somewhat larger than for the aPM. This is the result of a too high proton density close to the sources in the absence of convection. For the other regions the γ-ray flux in the aPM is insignificantly larger than in the isotropic model, which is mainly the result of slight deviations in the propagated proton and electron spectra. The latitude and longitude profiles for the inner Galaxy are presented in Figs. 4.15 and 4.16 for 100 MeV $< E <$ 500 MeV. The model shows the same deficiency of IC emission at intermediate latitudes as the conventional isotropic model, this can also be seen from region D in Fig. 4.14. Note, that we use a simple power-law extragalactic background model with $I_{EB} \propto E^{-2.1}$. A more detailed model may yield a better description of the γ-ray flux from the halo and at intermediate latitudes. However, this extragalactic background model should be determined from the Fermi-LAT data. The longitude distribution at intermediate latitudes, shown in Fig. 4.17, reveals that the deficiency almost exclusively results from a lack of diffuse γ-rays from the GC region for both, the aPM and the isotropic model. Since bremsstrahlung and emission from π^0-decay are rather flat for intermediate latitudes, an increased contribution from IC would improve the model prediction. Independent of direction the model prediction below 100 MeV is somewhat too low, which also indicates that the contribution from IC is underestimated. The deficiency in IC can be remedied by assuming that the Galactic electron density is somewhat larger than the local electron density, thus leading to a larger contribution from IC at low energies, as has been done previously by Strong et al. (2004a). Figure 4.18 shows the γ-ray spectra for a Galactic electron density increased by a factor 1.5, while the Galactic proton density remains unchanged. Figures 4.19 and 4.20 show the longitude and latitude profiles for the inner Galaxy for an aPM with the Galactic electron density increased by a factor 1.5 and the proton density kept constant. The right side of Fig. 4.17 shows the longitude profile between 150 and 300 MeV for intermediate latitudes. Whether or not the Galactic electron density can be different from the local electron density strongly depends on the transport parameters and the electron energy losses. Another possibility would be an untraced component of non-equilibrium gas (Breitschwerdt & de Avillez, 2006). In this case an additional bremsstrahlung contribution from electrons in the hot gas would be expected. Such a contribution would be most significant at large latitudes in the source region, where the relative component of the hot gas blown out by SNs is expected to dominate. This is exactly what is required by the latitude and longitude profiles in Figs. 4.15 and 4.17.

Alternatively an increase in the proton density above the plane in the inner Galaxy would lead to a more pronounced peak in the π^0 emission at intermediate latitudes. This could be achieved by a smaller diffusion coefficient in this region which would be well motivated by the decrease of the regular magnetic field above the disk. However, this will not improve the model prediction below 100 MeV, so additionally a softer electron spectrum would be required.

As discussed earlier, the aPM is almost completely independent of the position of the boundary condition, i.e. the halo size, provided, that the boundary is positioned well outside the diffusion zone limited by z_c. In this model it is therefore possible to examine the impact of additional IC emission from an extended halo with z_h=100 kpc without loosening the constraints on local CRs. The gain in additional IC at intermediate latitudes is at the % level and thus negligible in the total flux. An increased halo therefore will not improve the γ-ray emission from the halo significantly. An additional untraced gas component in the source region, a larger Galactic electron density or possibly an additional electron population appears to be required both in the isotropic model and the aPM.

4.4.2 Soft γ-ray Gradient

The aPM, as well as the isotropic models, require a somewhat increased Galactic electron density. In the isotropic case this leads to a pronounced peak in the Galactic IC emission from the GC, as can be seen from the top row in Fig. 4.21. As a possible solution a significantly flatter source distribution has been proposed by Strong & Moskalenko (1998), shown in Fig. 4.2 as the black full line, which reduces the contribution from π^0-decay and bremsstrahlung in the GC. However, this source distribution has to be chosen *ad hoc* according to what is expected from γ-rays. An increase in the X_{CO} scaling factor towards

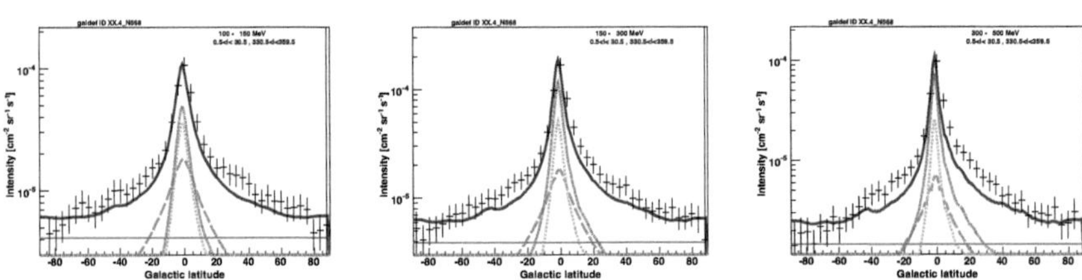

Figure 4.15: Latitude profiles for the inner Galaxy ($|l| < 30.5$) for the aPM compared to the EGRET data between 100 and 500 MeV.The EGRET-excess above 500 MeV is not confirmed by preliminary FERMI data, so only data below 500 MeV are considered here. Line coding as in Fig. 4.14

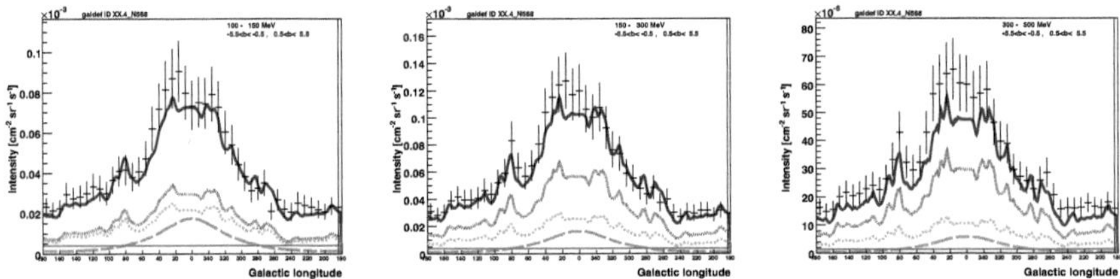

Figure 4.16: Longitude profiles for the Galactic disk ($|b| < 5.5$) for the aPM compared to the EGRET data between 100 and 500 MeV. Line coding as in Fig. 4.14.

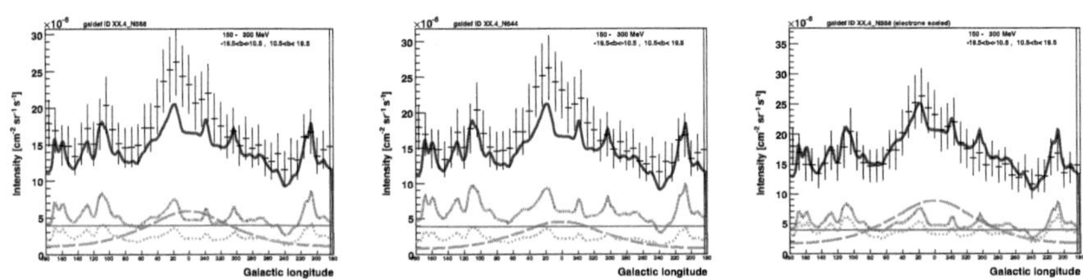

Figure 4.17: Longitude profiles for region D in the aPM (**left**), an isotropic model (**middle**) and the aPM with the Galactic electron density increased by a factor 1.5 (**right**) for γ-rays between 150 MeV and 300 MeV. The full blue line is the sum of the contributions from inverse Compton (green dashed), bremsstrahlung (light blue dotted), π^0-decay (red fine-dotted) and extragalactic background (purple full).

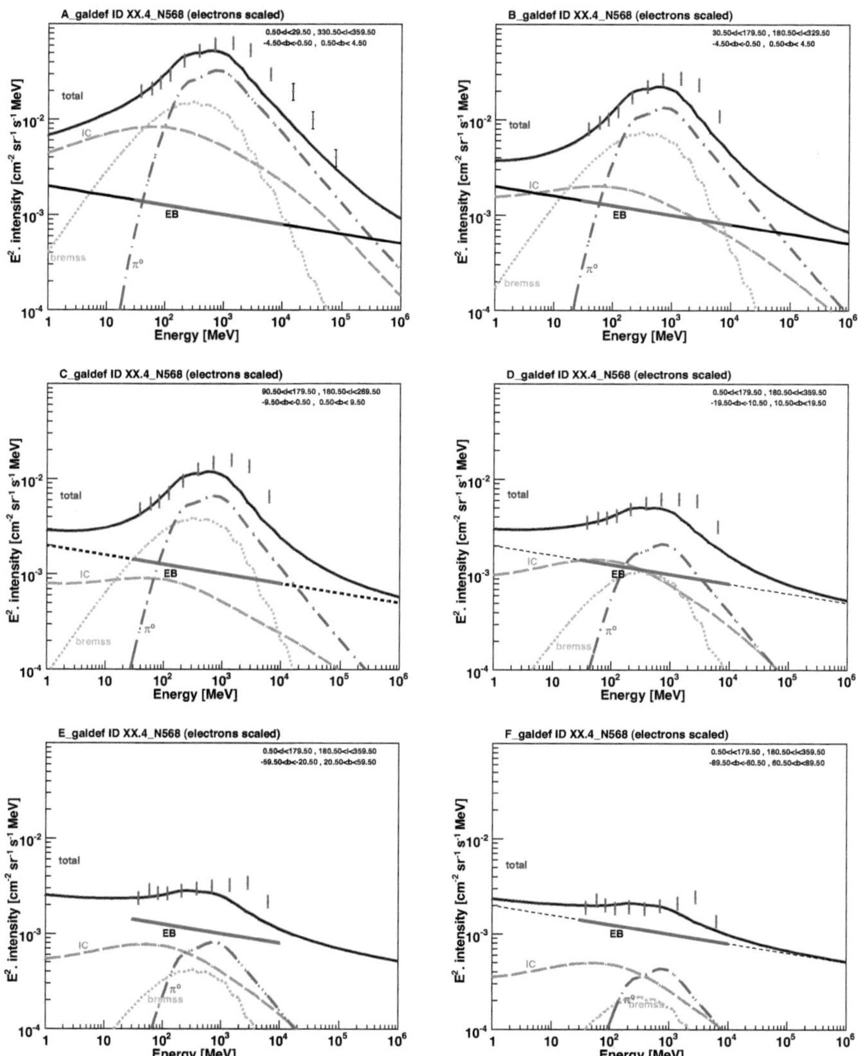

Figure 4.18: Diffuse γ-rays for the six different sky regions as defined in Strong et al. (2004a) for an aPM with the Galactic electron density increased by a factor 1.5. Line coding: bremsstrahlung (*light blue dashed*), inverse Compton (*green long dashed*), π^0-decay (*red long dashed-dotted*), total (*blue full*).The pink full line is the extragalactic background model according to Sreekumar et al. (1998). The EGRET data are corrected for the PSF.

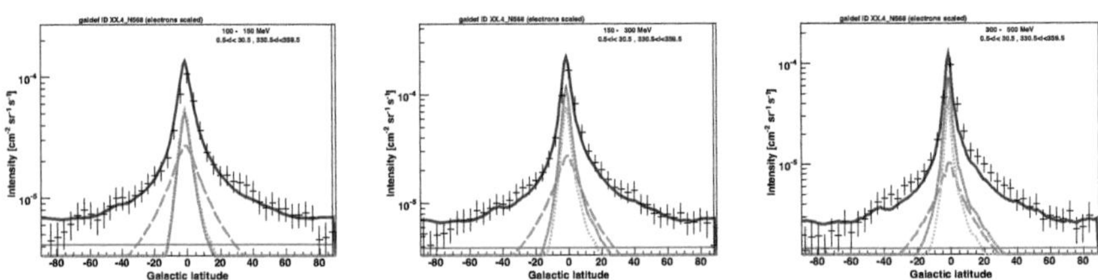

Figure 4.19: Latitude profiles for the inner Galaxy ($|l| < 30.5$) for the aPM with the Galactic electron density increased by a factor 1.5 compared to EGRET data between 100 and 500 MeV. Line coding as in Fig. 4.18.

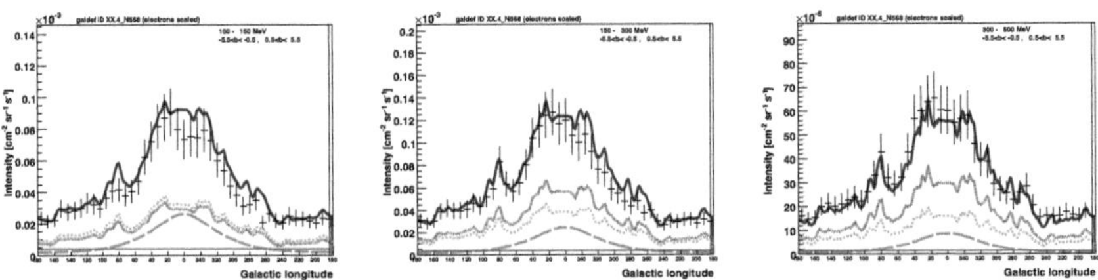

Figure 4.20: Longitude profiles for the Galactic disk ($|b| < 5.5$) for the aPM with the Galactic electron density increased by a factor 1.5 compared to EGRET data between 100 and 500 MeV. Line coding as in Fig. 4.18.

the outer Galaxy is a more likely explanation, but the gradients required by the EGRET data are on the limit of what is expected from the increase in metallicity (Strong et al., 2004b).

As mentioned previously the self-consistent Galactic wind calculations by Breitschwerdt et al. (2002) predict a softer γ-ray gradient, because the CR escape time varies with Galactocentric radius depending on the local source strength. Enhanced particle injection by the sources therefore results in enhanced CR escape and thus smoothens the propagated CR distribution. Since γ-rays from π^0-decay predominantly originate from GeV protons it is interesting to compare the radial proton distribution at this energy. Figure 4.22 shows the radial distribution for 1-5 GeV protons in an aPM and an isotropic model with $z_h = 4$ kpc, both models use the SNR distribution as the source distribution. In the isotropic model the propagated proton distribution still resembles the strong peak of the source distribution, which will lead to problems in the γ-ray production rate unless a strong increase in X_{CO} is assumed. Protons in an aPM are significantly flatter than protons in the isotropic model and a constant X_{CO} scaling factor is almost consistent with the observed flat γ-ray gradient, even if the Galactic electron density is increased to match the IC emission from intermediate latitudes. The bottom row of Fig. 4.21 shows the longitude profile for the Galactic disk in this case. The improvement due to the flatter profile of emission from π^0-decay and bremsstrahlung is clearly visible, but above 100 MeV a slight excess in emission from π^0-decay and bremsstrahlung is still visible. A fine-tuned radial dependence of the convection velocity or a rather soft gradient in X_{CO} appears to be compatible with the data.

4.4.3 Radio Emission in an aPM

We checked that the electron energy losses via synchrotron radiation are reasonably well described by calculating the radio emission in longitude and latitude in comparison to the 408 MHz data from the (Haslam et al., 1982) sky map.

Following Moskalenko et al. (1998) we use the following parameterization of the total regular magnetic field for the calculation of the electron energy losses:

$$B_{reg}(R) = B_0 e^{[\frac{-R-R_0}{R_B}]} e^{[-\frac{|z|}{z_B}]}, \tag{4.5}$$

with $R_B = 10$ kpc and $z_B = 0.2$ kpc. Figure 4.23 shows the latitude and longitude profile of synchrotron radiation in an aPM at 408 MHz and the synchrotron spectrum. We choose $B_0 = 6.5\ \mu$G in order to best reproduce the (Haslam et al., 1982) all-sky map. The spatial shape of the model prediction is not too good, which is due to a too simple parameterization of the magnetic field. However, here we are not interested in a good synchrotron prediction, but rather use the synchrotron data to constrain the magnetic field strength. This magnetic field strength is used in order to calculate the electron and positron synchrotron losses during propagation. A choice of B_0 which, on the average, reproduces the observed synchrotron radiation after propagation ensures a consistent estimate of the electron and positron synchrotron losses.

4.5 Interlude

In this chapter a new model for Galactic cosmic ray transport was presented, which allows for significant convective transport as expected from the Galactic winds deduced from

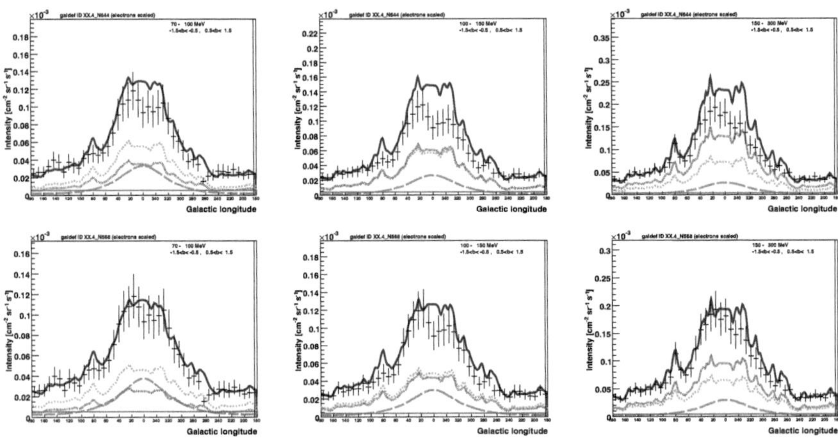

Figure 4.21: Longitude profile for the disk ($1.5 \geq |b|$) for an isotropic model (**top row**) and an aPM (**bottom row**). For both models the Galactic electron density is increased by a factor 1.5 to meet the requirements from γ-rays at intermediate latitudes.

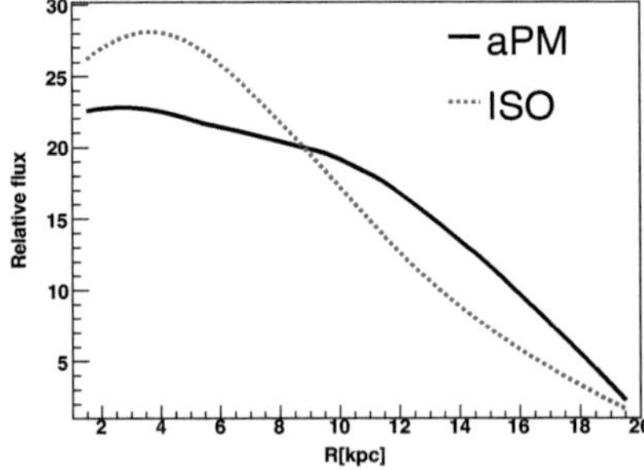

Figure 4.22: Radial distribution of GeV protons for the aPM and an isotropic model. The source distribution is in both cases that of (Case & Bhattacharya, 1996).

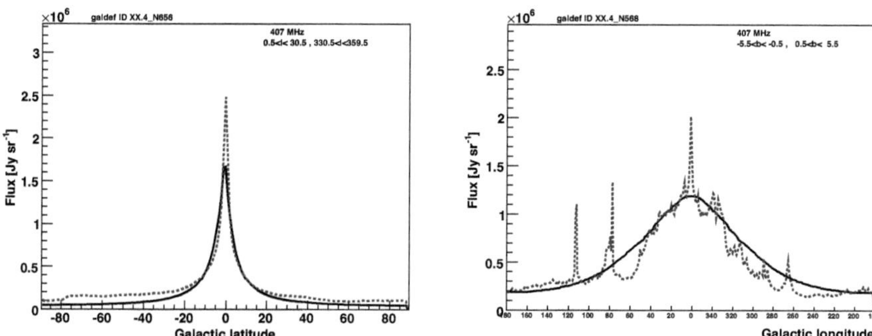

Figure 4.23: Synchrotron latitude profile (**top**) and synchrotron longitude profile (**bottom**) at 408 MHz an an aPM. The synchrotron data are used as a cross check for the electron and positron energy losses. A proper choice of B_0 according to the synchrotron data ensures a correct estimate of the electron and positron synchrotron losses during propagation. The spatial shape is not well reproduced due to our simple magnetic field model. Data (blue dotted): Haslam et al. (1982).

the X-ray data from the ROSAT satellite. The model has been realized by modifying the publicly available GALPROP code, which up to now allowed only isotropic transport. Isotropic transport models, which feature globally constant transport parameters, can only accommodate negligible convection speeds, since with convection the CRs are driven away from the disc, thus not returning often enough to the disk to produce secondary particles. A possible way to allow for large convection is to allow the diffusion in the halo to be different from the diffusion in the disk, i.e. to give up isotropic propagation with globally constant transport parameters. The GALPROP code was modified in the following way:

- the Galactic winds were assumed to be proportional to the CR source distribution, which was taken to be the SNR distribution

- the mean free path of CRs - and therefore the diffusion coefficient - in the halo was assumed to increase linearly with the distance from the disk. Although the exact spatial dependence of a diffusion coefficient generated by CRs leaving the Galaxy can be complicated (see e.g. (Dogiel & Gurevich, 1993) and (Dogiel et al., 1994) for a derivation of the spectrum of turbulences in the halo), the locally measured isotropy of CRs suggest a larger diffusion coefficient in the halo (Ptuskin & Khazan, 1976).

Fixing the magnitude of the convection speed to the wind speeds suggested by the ROSAT data, the increase in diffusion coefficient in the halo can be fitted from the amount of secondary production (from B/C ratio) and the residence time of CRs (from the cosmic clocks, in this case the $^{10}Be/^{9}Be$ ratio). It is shown that such a model is consistent with all available CR data, including not only the ROSAT data on convective winds, but also the large bulge/disk ratio of the positron annihilation line as observed by the INTEGRAL satellite. In the anisotropic model presented here the small positron annihilation signal from the disk can be explained by fast escape of positrons from the disk. The large B/D ratio can be naturally explained by the energy independent convective transport of low energy positrons from the disk to the halo, where there are no electrons to annihilate with. Convective transport is absent in the bulge, because the gravitational potential is too strong there to launch Galactic winds.

An additional interesting feature of the present model is the smooth transition to free escape of CRs, because of the increase in mean free path with increasing distances from the disk. Therefore the boundary condition can be moved to infinity in contrast to isotropic propagation models, where the boundary condition is fine-tuned to get the correct residence time of CRs inside the Galaxy.

The model describes the local fluxes and relative abundances of charged CRs well, a cross check with the published EGRET data on diffuse γ-rays below 1 GeV reveals the same deficiency as the widely-accepted isotropic models. For energies above 1 GeV a meaningful comparison to the data is difficult, since the preliminary Fermi-LAT data on diffuse γ-rays do not confirm the EGRET excess. Furthermore, the modified GALPROP version we developed is capable of simulating detailed structures in the ISM and their respective counterparts in the transport parameters. Thus, the code developed in this work will be a valuable tool to examine the details of CR transport that will become accessible to us with the upcoming data by Fermi, PAMELA and AMS-02.

In the next chapter we will address the issues discussed above: On the one hand we will compare our model predictions to the preliminary *unpublished* Fermi-LAT data on diffuse γ-rays, exercising due care with respect to their preliminary nature, and on the other hand we will discuss some of the small scale features of our Galaxy, that are expected to have an impact on CR transport, such as local deviations in the gas density or the spiral structure of the Milky Way, in the context of the aPM.

The next chapter is called "The Dark Chapter" for two reasons: Firstly, we will address indirect dark matter searches in diffuse γ-rays and charged CRs and, secondly, we will discuss the current dark spots in CR transport modelling, which are basically given by the limited knowledge of the detailed geometry and the magnetic fields of the Milky Way.

Chapter 5

The Dark Chapter

The universe is a lot more complicated than you might think even if you start from a position of thinking that its pretty damn complicated to begin with.

Douglas Adams - Mostly Harmless

The dark matter (DM) problem is fascinating scientists for more than 75 years now. A variety of observations, from galactic to cosmological scales, lead to the conclusion that an unknown form of matter must exist which contributes significantly to the energy density in the Universe. Early observational evidence was given by Zwicky's observation of a large velocity dispersion of the Coma Cluster (Zwicky, 1933) and, a few years later, Babcock's measurements of the fast rotation of the stars in the Andromeda galaxy (Babcock, 1939). Both these measurements indicate too large velocities to be bound by Newtonian gravity, so that an additional invisible matter component has to be assumed.

Today, an impressive amount of data from studies of the microwave background radiation, supernova distance measurements, and large-scale galaxy surveys have solidified the Standard Model of cosmology. In this model structure formed through gravitational amplification of small density perturbations with the help of cold dark matter. Without the existence of dark matter the density contrast seen in the universe today could not have formed, given the small amplitude of density fluctuations inferred from anisotropies of the cosmic microwave background. Especially the WMAP measurements (Komatsu et al., 2009) of the fluctuation in the CMB radiation allow us to make precise predictions about the properties and abundance of this unknown matter. The relative height and positions of the first few peaks of the multipole spectrum of CMB tell us that dark matter makes up about 22% of the energy of the universe and that it must be non-baryonic. The baryonic constituents of matter only add up about 4% of the total energy, the remaining energy is believed to occur in the form of dark energy.

Recently, the PAMELA (Adriani et al., 2009a,b), Fermi (Abdo et al., 2009) and ATIC (Chang et al., 2008) results on electrons and positrons have led to an explosion of papers, interpreting these observations as a signal of dark matter annihilation (DMA) (or decay). In the following we will discuss some contemporary DMA interpretations of astrophysical observations and comment on the astrophysical uncertainties in the background which is given by CRs. After introducing the most important particle DM candidates in Section 5.1, we will focus on γ-rays in Section 5.2 and discuss the DMA interpretation of the EGRET excess. Then we will turn to the preliminary diffuse Fermi-LAT data and discuss whether or not a DM candidate as expected from EGRET (de Boer et al., 2005) is compatible with the Fermi-LAT data. Unlike the EGRET data, the Fermi-LAT data makes an astrophysical interpretation plausible and we will discuss a model with hard electron and proton spectrum to explain the FERMI results. We will see that with or without an additional contribution from DM, the Fermi data require a different normalization for diffuse γ-rays than for the local electrons and protons. We will discuss possible causes of this discrepancy in Section 5.3. We will then turn to the local antiproton flux which also constitutes a problem for most CR transport models: while conventional CRs within GALPROP can only account for 60% of the observed antiprotons, the additional contribution from DMA overshoots the data by an order of magnitude (for the case of a 60 GeV neutralino as expected from EGRET). We will discuss the uncertainties in the local antiproton flux from CRs and DMA in Section 5.4. Finally, in Section 5.5, we will review some contemporary DMA interpretations and comment on the uncertainties from CR transport within the context of the aPM.

5.1 Indirect Dark Matter Searches and Dark Matter Candidates

The dominant fraction of dark matter has to be non-baryonic and only weakly interacting. It must be stable on cosmological timescales or it would have disappeared from the cosmic stage long ago. If the standard concept of quantum field theory is used to describe the properties of elementary particle candidates, as it is in most current models of dark matter,

Table 5.1: Properties of various Dark Matter Candidates. Adopted from Bergström (2009).

Type	Particle Spin	Approximate Mass Scale
Axion	0	μeV-meV
Inert Higgs Doublet	0	50 GeV
Sterile Neutrino	1/2	keV
Neutralino	1/2	10 GeV - 10 TeV
Kaluza-Klein UED	1	TeV

the candidate particle can be characterized by the mass and spin. The mass of proposed candidates spans a very large range, as illustrated in Table 5.1.

The averaged density of cold dark matter (CDM) is now known to an accuracy of a few percent. With h being the Hubble constant today in units of 100 $\mathrm{kms^{-1}Mpc^{-1}}$, the density derived from the 5-year WMAP data (Komatsu et al., 2009) is

$$\Omega_{\mathrm{CDM}}h^2 = 0.1131 \pm 0.0034, \tag{5.1}$$

with the estimate of $h = 0.705 \pm 0.0134$.

Assuming thermally produced dark matter this corresponds to a cross section averaged over velocities at the time of thermal decoupling of (Jungman et al., 1996)

$$\langle \sigma_A v \rangle = 2.8 \cdot 10^{-26}\ \mathrm{cm^3 s^{-1}}. \tag{5.2}$$

Interestingly, this averaged cross section just corresponds to what one gets with a weak interaction cross section for particles of mass around the electroweak scale of a few hundred GeV and this agreement is sometimes called the "WIMP miracle" (WIMP standing for Weakly Interacting Massive Particle). Of course the "WIMP miracle"may be coincidence, but most of the detailed present-day models proposed for the dark matter are in fact containing WIMPs as dark matter particles.

When estimating the observable annihilation rate today, one has to keep in mind that there are large astrophysical uncertainties arising from the presence of substructure in the dark matter distribution. Such substructures have been discovered in large simulations of structure formation (Diemand et al., 2007; Springel et al., 2008).

For the indirect detection of DM diffuse γ-rays are a good tracer of the DM distribution, because they are not bend by the Galaxy's magnetic field and suffer only insignificant energy losses.

The DMA signal depends on the distribution of WIMPs. The annihilation rate is proportional to the number densities of particles and of possible annihilation partners which in this case are the same species, i.e. the signal is proportional to the squared number density. To take the DM density profile into account one has to integrate along the line of sight to get the total flux in a particular direction (Bergström, 2009):

$$\Phi_{\rm dm}(E,\psi,\Delta\Omega) = \frac{\langle\sigma v\rangle}{4\pi}\cdot\sum_f \frac{dN}{dE}b_f\cdot\frac{1}{\Delta\Omega}\int_{\Delta\Omega}\int_{\text{line of sight}}\frac{1}{2}\frac{\rho_\chi(l)^2}{m_\chi^2}dl_\psi \tag{5.3}$$

Here Φ is the differential flux at the energy E in a direction ψ for a cone with a solid angle $\Delta\Omega$; $\langle\sigma v\rangle$ is the total thermal averaged annihilation cross section; the factor b_f is the fraction of a particular final state with the differential number of photons per annihilation at the given energy dN/dE; the number density of the WIMPs is given by ρ_χ/m_χ, the factor 1/2 is due to the fact the dark matter constitutes its own antiparticle.

A fraction of the WIMPS is expected to be distributed in small scale clumps with masses above 10^{-6} $M_\odot$ (Berezinsky et al., 2008) as a result of hierarchical clustering. Any deviation from a smooth distribution of the WIMPs results in an enhanced signal, because the annihilation rate is proportional to $\rho_\chi^2(l)$. The procedure in this situation has been to introduce a "boost factor". If the clumps are distributed in a way that the median density of the clumps is proportional to the local median density, then the median squared density is proportional to the square of the local median density. Such a model can be described as a halo of amorphous interpenetrating substructures. The boost factor in this case is dimensionless.

$$\langle\rho_\chi^2\rangle \approx b\cdot\langle\rho_\chi\rangle^2 \tag{5.4}$$

Unfortunately the boost factor does not have a unique definition in the literature. Many authors do not boost the annihilation rate as in Eq. 5.4, but the detections rates (see e.g. Bergström (2009)). This leads to large uncertainties in the predicted signal, especially when it comes to charged annihilation products. For these the effects of propagation then have to be invoked in the boost factor. For example, a nearby clump of DM, would lead to a larger boost factor for protons than for electrons, simply due to the large electron energy losses. Any definition of the boost factor in the form of an increase in detection rate is forced to make implicit assumptions about the transport model in addition to the clumpiness of DM. This sometimes makes the fits of the charged annihilation products of a certain DM candidate hard to interpret, because usually a more or less detailed transport model is used to calculate the background from CRs. Boost factors applied to the detection rate implicitly depend on particle type and energy and therefor are only indirectly linked

to the clumpiness of DM.
Since we have a detailed transport model at hand we will use the boost factor as defined in Eq. 5.4, i.e. the boost factor is a measure for the clumpiness of DM. This means that the boost factor is applied to the annihilation rate, and *not* to the detection rate.
For γ-ray observations, the enhancement should be computed within the line of sight cone, and therefore one may get different boost factors for different directions.
The computation of the boost factor in realistic astrophysical and particle physics scenarios is a challenging task, which has so far only been partially addressed (see e.g. Berezinsky et al., 2008) in analytical calculations. N-body simulations do not have the resolution to treat clumps as small as 10^{-6} solar masses.
Among the candidates for dark matter currently under discussion are non-particle candidates, such as primordial black holes and a variety of particle candidates, such as axions, inert higgs, sterile neutrinos (which because of their large mass would be warm dark matter), and WIMPS like the lightest supersymmetric particle and the lightest Kaluza-Klein particle.
Recently the PAMELA positron fraction and the ATIC and Fermi results on electrons, created a blast of papers with dark matter interpretations. Among these the axion, the lightest supersymmetric particle (LSP) and the lightest Kaluza-Klein particle (LKP) have been used to explain these observations as a signal from dark matter annihilation (or decay). Although this work focusses on CR transport and does not deal with the nature of dark matter, these dark matter candidates are of special interest to us in the sense that their stable, charged decay products are subject to conventional CR propagation. Some of them are therefore introduced in brevity.

Standard model neutrinos only interact by the weak force and they are known to be massive from the observation of neutrino oscillations. They have been considered as candidates for dark matter in the past and among the candidates presented here, they are special because their existence is not hypothetical but has been well established. Their relic density can be calculated to be

$$\Omega_\nu h^2 = \frac{\sum m_\nu}{93\,\mathrm{eV}} \tag{5.5}$$

Due to their low mass, neutrinos are relativistic and therefore a candidate for so-called hot dark matter. Data on the large-scale structure of the Universe, combined with anisotropies in the cosmic microwave background and other cosmological probes can be used to set an upper limit of 0.17 eV (95 % confidence level) on the neutrino masses (Seljak et al., 2006), implying a relic density of not more than $\Omega_\nu h^2 < 0.006$, which is not enough for neutrinos to be the dominant form of dark matter.

Axions are one of the earliest suggestions for particle dark matter. The axion has initially been introduced to solve the strong CP-problem (Peccei & Quinn, 1977; Weinberg, 1978; Wilczek, 1978). In general, the action density of the standard model includes a term

$$\mathcal{L}_{\rm SM} = \ldots + \frac{\theta g^2}{32\pi^2} G^a_{\mu\nu} G^{a\mu\nu} \tag{5.6}$$

where $G^a_{\mu\nu}$ are the QCD field strengths, g is the QCD coupling constant, and θ is a parameter. The observed physics depends on the value $\bar{\theta} \equiv \theta - \arg\det m_q$ where m_q is the quark mass matrix. While the term in (5.6) violates the C and CP symmetries, as do the weak interactions in the standard model, the experimental upper bound on the electric dipole moment of the neutron limits $|\bar{\theta}| < 10^{-10}$ (Asztalos et al., 2006) and the question arises why $\bar{\theta}$ is so small when it can be expected to be an arbitrary number. It was shown that the introduction of an additional field $A(x)$, called the axion, can naturally explain why $\bar{\theta}$ is zero. The corresponding term in the action is

$$\mathcal{L}_{\rm axion} = \frac{1}{2}\partial_\mu A \partial^\mu A + \frac{g^2}{32\pi^2}\frac{A(x)}{f_A} G^a_{\mu\nu} G^{a\mu\nu} \tag{5.7}$$

f_A is a constant with dimension of energy, and the mass and couplings of the axion can be expressed in terms of this constant, $m_A, g_{Aii} \propto f_A^{-1}$. The allowed axion mass range is limited from below by cosmological bounds and from above by the physics of stellar evolution and SN dynamics to lie in the range $10^{-6} \sim 10^{-3}$ eV. Nevertheless, the axion is a viable candidate for cold dark matter, with relic density $\Omega_A \propto m_A^{-7/6}$, because cold, non-thermal axions may have been produced during the QCD phase transition in the early Universe, so they could have been produced in large numbers out of thermal equilibrium. Searches for cosmological and solar axions are underway, but they have eluded discovery so far.

Lightest Supersymmetric Particle arises in the context of supersymmetric extensions of the standard model of particle physics (Ferrara et al., 1974; Wess & Zumino, 1974a,b). Although the standard model of particle physics has been enormously successful in describing the interactions of matter at the most fundamental level, it has a number of theoretical shortcomings. Two examples are the hierarchy problem and the problem of unification of the gauge couplings. The former is related to the question why the Higgs mass is so small. While the mass scale of the standard model is set by the vacuum expectation value of the Higgs $v \approx 246\,\text{GeV}$, divergent quadratic loop corrections to the Higgs mass occur, $\delta m_H^2 \sim \Lambda^2$, where Λ is a cut-off scale at which the standard model must be modified to remain valid. This is usually associated with the Planck scale, $M_P = (G_N)^{-1/2} \approx 1.2 \cdot 10^{19}\,\text{GeV}$, which means that the mass parameter μ in the Higgs potential $V = -\mu^2\phi^\dagger\phi + \lambda/4(\phi^\dagger\phi)^2$ must be of a similar amplitude to cancel the divergence. This large fine-tuning, where two large mass scales almost cancel to produce the observed masses of the standard model, seems unnatural and is known as the hierarchy problem. The second example revolves around the unification of gauge couplings. The running of the gauge couplings in the standard model as a function of the energy scale is described by the renormalization group equations (RGEs). The inverse gauge couplings $\alpha_1^{-1}(Q^2)$, $\alpha_2^{-1}(Q^2)$, and $\alpha_3^{-1}(Q^2)$ fail to meet at high energies, although they come close to doing so. A unification of the gauge couplings is a highly desirable property of a fundamental theory. In fact, these and other problems can be overcome in supersymmetric extensions to the standard model (Amaldi et al., 1991; Bertone et al., 2005; de Boer, 1994; Tata, 1997).

To prevent rapid proton decay a conserved multiplicative quantum number, called R-parity, is introduced:

$$R = (-1)^{3(B-L)+2S}\,, \tag{5.8}$$

where B is the baryon number, L the lepton number and S the spin of the particle. This implies that $R = +1$ for ordinary particles and $R = -1$ for supersymmetric particles. If R-parity is conserved supersymmetric particles can only be created or annihilate in pairs in reactions of ordinary particles. This way a single supersymmetric particle can only decay into final states containing an odd number of supersymmetric particles. As a consequence the lightest supersymmetric particle is stable, since there is no kinematically allowed state with negative R-parity which it can decay to. In the context of supersymmetry the neutralino χ is the most discussed candidate. If neutralinos are indeed the particle constituting most of the dark matter, then in the early universe neutralinos χ were created by pair-production. As the universe kept cooling they left thermal equilibrium and at some point the expansion rate of the universe exceeded the annihilation rate which lead to a non-zero relic abundance today. If the scale of supersymmetry breaking is related to that of electroweak breaking, χ will be a WIMP and Ω_χ will be of the right order of magnitude to explain the non-baryonic cold dark matter. Even if the particle content of the supersymmetric model is kept minimal (as in the Minimal Supersymmetric Model MSSM) the number of free parameters is unhandily large (of the order of 100, although unhandyness is not an argument against the existence of these parameters). Most studies have focused on constrained supersymmetric models, such as minimal supergravity (mSUGRA) models (Chamseddine et al., 1982), where one unifies not only gauge couplings, but also susy breaking terms. In such models the total number of parameters is reduced to 5 or 4 plus a sign of the Higgsino mass parameter (for a detailed discussion of dark matter in mSUGRA models see Edsjo et al. (2003)).

Lightest Kaluza-Klein particles arise in theories of universal extra dimensions. Here it is assumed that there are dimensions in addition to the known four-dimensional space-time. The additional dimensions have not been observed yet, so they have to be compactified which introduces some characteristic scale R. This leads to the appearance of a so-called tower of new particle states in the effective four-dimensional theory, with the mass of the n-th Kaluza-Klein (KK) mode given by

$$m^{(n)} = \sqrt{(n/R)^2 + m^2} \tag{5.9}$$

for a standard-model particle of mass m. Assuming a symmetry called KK parity, the lightest KK state (LKP) can be stable and therefore constitutes a candidate for dark matter (Kolb & Slansky, 1984; Servant & Tait, 2003). It is likely to be associated with the first excitation of the hypercharge gauge boson, the $B^{(1)}$. If its mass is on the order of 1000 GeV the LKP can explain the observed relic density Ω_{dm}. Recently the $B^{(1)}$ has attracted a significant amount of attention, because it has the attractive feature of dominantly producing charged leptons in its annihilation. This would provide a source of hard positrons and electrons as indicated by the recent PAMELA and Fermi measurements.

Other candidates Many other candidates including sterile neutrinos, gravitinos and little Higgs models have been proposed (Bertone et al., 2005). Positrons from annihilation of light scalar dark matter have been proposed to cause the 511 keV-line observed in the direction of the Galactic bulge. Little Higgs models, introduced as an alternative mechanism to supersymmetry to stabilize the weak scale, may contain a dark matter candidate. Superheavy dark matter particles, so called wimpzillas, would be interesting because of their expected contribution to ultra-high energy CRs.

5.2 Diffuse Galactic γ-rays: EGRET and Fermi-LAT, Dark Matter and Astrophysics

Here we will review the DMA interpretation of the EGRET excess and compare the DM model derived from the EGRET data to the preliminary Fermi-LAT data in diffuse γ-rays. We will also discuss possible astrophysical explanations of the data, which, given the generally softer Fermi spectrum, appear more likely than in the case of EGRET.

5.2.1 The EGRET γ-ray Excess

As discussed earlier an excess of diffuse gamma rays has been observed by the EGRET telescope on board of NASA's CGRO (Compton Gamma Ray Observatory) (Hunter et al., 1997). Below 1 GeV the CR interactions describe the data well, but above 1 GeV the data are up to a factor two above the expected background (see e.g. Fig. 4.14). This so-called EGRET excess was analyzed using a data-driven calibration of the background (de Boer et al., 2005). Such an approach is particularly suitable if the shape of the dominant background is well known. For diffuse Galactic gamma rays this is π^0 production in inelastic collisions of CR protons on the hydrogen gas of the disk, as we have seen in Section 4.4.1. The shape of the resulting γ-ray spectrum is known from fixed target accelerator experiments and the shape of the DMA signal is known from e^+e^- annihilation. Since the signal has a significantly harder spectrum than the background one can perform a data-driven analysis by fitting the two shapes to the experimental data with a free normalization for each shape, thus obtaining the absolute contribution from signal and background for each sky direction in a rather model-independent way. This analysis assumes that the uncertainties in the interstellar background shape, i.e. the proton and electron spectrum are small. A simultaneous fit of DMA signal and background shape has been made to 180 independent sky directions. The average χ^2 per degree of freedom summed over all (ca. 1400) data points is around 1, indicating that the errors are correctly estimated. The resulting background and signal is shown in Fig. 5.1 for the GC region. The excess is compatible with a WIMP mass between 50 and 70 GeV and the DMA interpretation of the EGRET excess is consistent with the expectations from Supersymmetry (de Boer et al., 2006).

Independent support for the DMA interpretation of the EGRET excess is delivered from a variety of measurements:

- The derived DM halo profile shows some unexpected substructure: outside the disk it corresponds to a cored halo profile, but inside the disk it reveals two additional doughnut-like structures at distances of about 4 and 13 kpc from the Galactic center (see Fig. 5.2). Structures like this are expected from the tidal disruption of dwarf galaxies captured in the gravitational field of our Galaxy. The "ghostly" ring of stars or Monocerus stream (with about $10^8 - 10^9$ solar masses in visible matter) could be the tidal streams of the Canis Major dwarf galaxy (see e.g. Martin et al. (2004); Peñarrubia et al. (2005) and references therein). If so, the tidal streams predicted from N-body simulations are perfectly consistent with the ring at 13 kpc (Peñarrubia et al., 2005). The ring at 4 kpc might also originate from the disruption of a smaller dwarf galaxy, but here the density of stars is too high to find evidence for tidal streams. However, direct evidence of a stronger gravitational potential well in this region comes from the ring of dust at this location. Since this ring is slightly tilted with respect to the plane its presence and orientation can be explained by the presence of a ringlike structure of DM.

- The halo profile derived from the EGRET data agrees with the rotation curve of the Milky Way as shown in Fig. 5.1. Note that the change of slope close to the position of the Sun is well described by the presence of the outer ring. However, one has to keep in mind that the rotation velocities in the inner and outer Galaxy are derived with different methods and, as always in astronomy, are subjects to artifacts.

- The strong gravitational potential well in the outer ring might cause the reduced gas flaring observed at the position of the ring (Kalberla et al., 2007). The half-width-half-maximum of the gas layer in the disk is shown on the right-hand panel of Fig. 5.1. The reduction in gas flaring corresponds to more than 10^{10} solar masses, in agreement with what one would expect from the EGRET data. It should be noted that the peculiar shape of the gas flaring was only understood after the astronomers heard about the EGRET ring. The effect is so large that visible matter cannot explain this peculiar shape, simply because there is not much visible matter beyond 10 kpc. A similar ring in the outer disk has been discovered in a nearby galaxy, indicating that such infalls may shape the disk and its warps (Martínez-Delgado et al., 2009).

The DMA interpretation of the EGRET excess received some serious criticism, e.g. the amount of dark matter would give rise to a too high Galactic surface density or the constraints from local antiprotons could not be met (Bergstrom et al., 2006). In the first

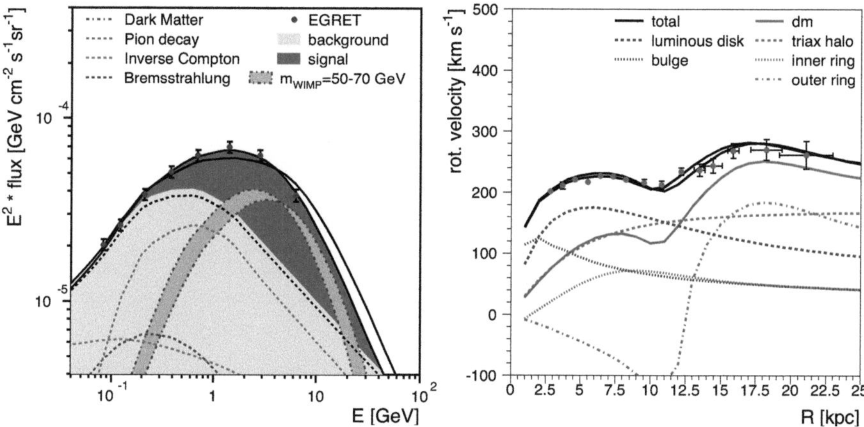

Figure 5.1: Left: Fit of the shapes of background and DMA signal to the EGRET data in the direction of the GC. The light shaded (yellow) area indicates the background using the shapes known from accelerator experiments, while the dark shaded (red) area corresponds to the signal contribution from DMA for a 60 GeV WIMP mass, where the small intermediate (blue) shaded area corresponds to a variation of the WIMP mass between 50 and 70 GeV. The two black full lines correspond to the sum of all contributions for a WIMP mass of 50 GeV (softer spectrum) and a WIMP mass of 70 GeV (harder spectrum). Taken from Ref. de Boer et al. (2005). **Right:** The Galactic rotation curve. Indicated are the different contributions from dark and baryonic matter. From de Boer et al. (2005).

case, one has to keep in mind that the diffuse halo component (which determines the gravitational effects of the DM halo) and the clumped halo component (which determines the annihilation signal), are not expected to have the same spatial distribution (Springel et al., 2008) and have been kept identical in the original analysis for reasons of simplicity. Bergstrom et al. (2006) also pointed out that the expected local antiproton flux from DMA would be factor of up to 100 too high if the charged stable decay products are taken into account. We will discuss antiprotons from DMA in Section 5.4 in detail.

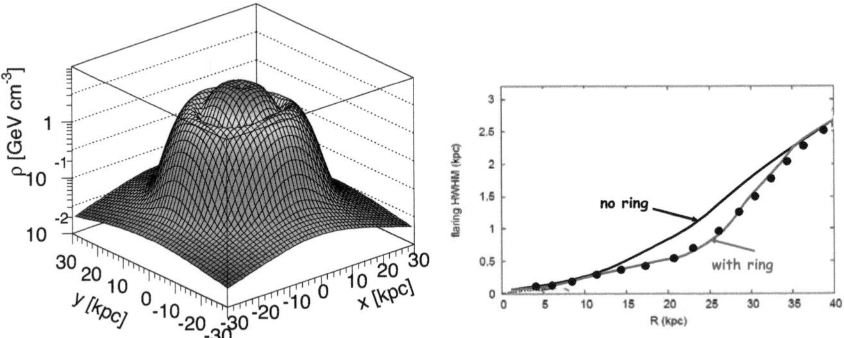

Figure 5.2: **Left:** The halo profile in the Galactic plane as derived from the EGRET data. From de Boer et al. (2005). **Right:** The half-width-half-maximum (HWHM) of the gas layer of the Galactic disk as function of the distance from the Galactic center. Clearly, the fit including a ring of dark matter above 10 kpc describes the data much better. Adapted from data in Kalberla et al. (2007).

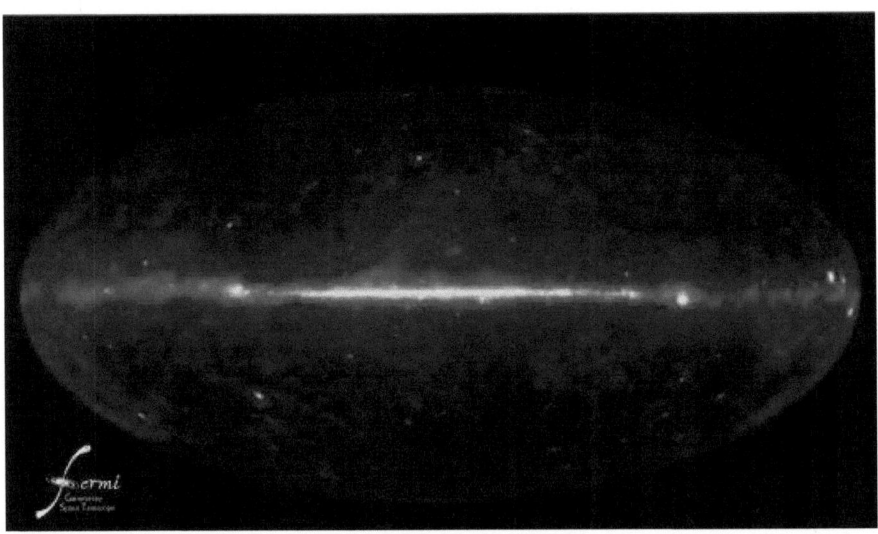

Figure 5.3: Fermi all-sky view after 1 year of data taking.

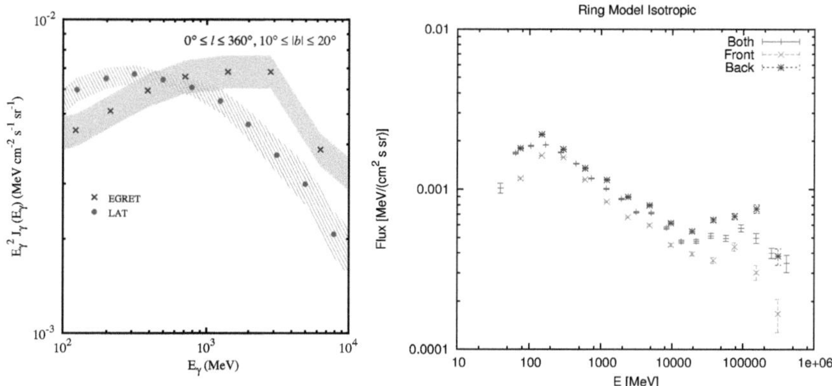

Figure 5.4: **Left:** Preliminary diffuse emission intensity averaged over all Galactic longitudes for latitude range $10° \le |b| \le 20°$ (region D) (Porter, 2009). Data points: LAT, red dots; EGRET, blue crosses. Systematic uncertainties: LAT, red; EGRET, blue. Point sources have not been subtracted. **Right:** Isotropic spectrum for the analysis of LAT data. Separate spectra are presented for Front and Back converting events because the residual charged particle background is different for them, the contamination by residual background being greater in the Back section. The 'Both' spectrum is the overall average as given in *isotropic_iem_v02.txt*. These are valid only for the P6_V3_DIFFUSE response functions and the *gll_iem_v02.fit* model of Galactic diffuse emission. Note that what is plotted in each case is E^2 times the differential intensity. **The broad feature near 100 GeV is understood to be due to residual (misclassified) heavy cosmic rays in the Pass 6 analysis and is not astrophysical.**

5.2.2 Fermi-LAT diffuse γ-ray model

The Fermi collaboration has recently made available data from the first year of data taking [1]. Figure 5.3 shows Fermi's all-sky view after one year of data taking. In most regions of the sky point sources like pulsars or SNRs contribute significantly to the γ-ray flux. An exception are intermediate latitudes. Here the diffuse emission is still significant, whereas only few point sources are present. Fig. 5.4 shows the "preliminary diffuse emission" (meaning no point sources have been subtracted) from intermediate latitudes compared to the EGRET diffuse γ-rays from this region. Clearly, the EGRET excess above 1 GeV is not confirmed by Fermi-LAT. In addition the Fermi-LAT data show a generally softer

[1]The data are publicly available from the Fermi Science Support Center at http://fermi.gsfc.nasa.gov/ssc

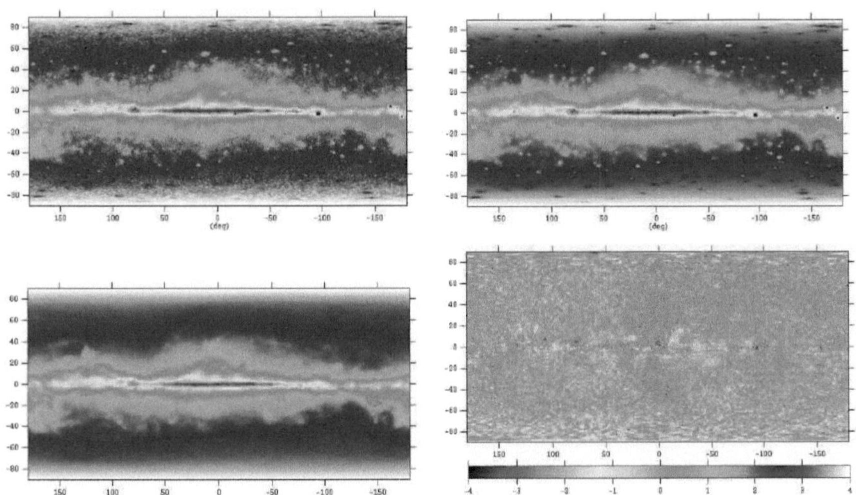

Figure 5.5: Top left: LAT all-sky γ-ray count map, $N_{obs}(l, b)$, in the 0.3-20 GeV energy band (log-scale). **Top right:** Diffuse model (*gll_iem_v02*) prediction together with modeled point sources, $N_{pred}(l, b)$, in the 0.3-20 GeV energy band. **Bottom left:** Diffuse model (*gll_iem_v02*) prediction alone, in the 0.3-20 GeV energy band. **Bottom right:** *Gll_iem_v02* residual map expressed in sigma values: $(N_{obs} - N_{pred})/\sqrt{N_{pred}}$. From Diffuse and Molecular Clouds Science Working Group Fermi-LAT (2009).

spectrum: even below 1 GeV the two measurements are incompatible.

Although a publication of the diffuse (meaning point source subtracted) emission is still pending the Fermi Science Support Center (FSSC) made available a preliminary model of the diffuse emission [2]. The *gll_iem_v02* model for the Galactic diffuse emission was developed using spectral line surveys of HI, CO (as a tracer of H_2) to derive the distribution of interstellar gas. Infrared tracers of dust column density were used to correct column densities in directions where the optical depth of HI was either over or underestimated. To allow for a Galactocentric gradient of cosmic-ray flux in the Galaxy, the $N(HI)$ column-densities and $W(CO)$ intensities have been derived for six Galactocentric rings. The model of the diffuse gamma-ray emission was then constructed by fitting the gamma-ray emissivities of the rings in several energy bands to the LAT observations. While the different gas column-density maps offer a template for photons originating from π^0-

[2]The diffuse background model is available from http://fermi.gsfc.nasa.gov/ssc/data/access/lat/BackgroundMc

decay and Bremsstrahlung emission, there is no simple template for the inverse-Compton emission I_{IC}. For this the prediction from GALPROP using the ISRF model from Porter (2005) was used.

Ten logarithmically distributed energy bands between 100 MeV and 10 GeV have been used to determine the differential emissivities $\frac{dq}{dE}$ for all the components. The spectral shapes of the emissivities were then frozen and, in a second step, the relative normalizations of the different contributions were obtained by fitting the single band γ-ray map integrated from 0.3 to 20 GeV.

For energies above 20 GeV the model has been extrapolated and globally renormalized to fit the LAT data so that the final model cube comprises 30 logarithmically-spaced energies between 50 MeV and 100 GeV. The isotropic component is not included in the cube, but is provided separately as a tabulated spectrum. The agreement between modelled emission and observed photon count is good as can be seen from the top row of Fig. 5.5, which shows the Fermi-LAT photon count map (top left) and the diffuse model *gll_iem_v02* plus modelled point sources (top right). The bottom row of Fig. 5.5 shows the diffuse model *gll_iem_v02* and the residual map for the *gll_iem_v02* model. Except for localized spots, where the point sources are not correctly modelled the deviation between model and data is on the 1 sigma level. Although this model is the most accurate the Fermi-LAT team has produced, it is accompanied by severe caveats (a detailed discussion can be found in Diffuse and Molecular Clouds Science Working Group Fermi-LAT (2009)), e.g.

- For energies above 20 GeV the model has been extrapolated and globally renormalized to the LAT data. A spectral fit has not been performed in this energy range.
- The diffuse model for energies below 120 MeV was not fitted to the data and is therefore less reliable.
- For energies above 50 GeV, the number of photons produced through interaction with the gas is probably underestimated compared to photons produced by IC scattering, so the spatial structure of the diffuse model above 50 GeV is probably too smooth.
- The diffuse model extends to 100 GeV in the current version, and the isotropic diffuse spectrum with residual background at higher energies was derived based on an extrapolation of that model. Studies of sources and diffuse emission at energies greater than 100 GeV are likely to be limited primarily by photon statistics but the reduced accuracy of the modeling at these energies should be kept in mind as well.

Given these constraints we limit our analysis to energies between 120 MeV and 20 GeV in the following.
The isotropic component was determined separately from the Galactic diffuse component by a maximum likelihood method including only high latitude emission ($|b| > 30^o$). It is continued to energies greater than 100 GeV using a simple extrapolation (linear in the logarithm of energy) of the diffuse Galactic emission model. *The isotropic component includes any true extragalactic component as well as charged particle background misreconstructed as γ-rays.* The right side of Fig. 5.4 shows the isotropic spectrum for the analysis of LAT data, which belongs to the diffuse emission model discussed above.
Here and in the following we are referring to the model *gll_iem_v02* from August 24, 2009 as "Fermi data". just in the same way we have been referring to the EGRET diffuse model as "EGRET data" up to now [3]. Note, however, that the model used here is a preliminary model. A new version is expected to be released in 2010. Despite the fact that we limit our analysis to the most reliable energy range, it should be kept in mind that the model prediction even in this range might change. More information about this preliminary diffuse emission model can be found in Diffuse and Molecular Clouds Science Working Group Fermi-LAT (2009).

Figure 5.6 shows the Fermi and EGRET data compared to the aPM. Since the diffuse Fermi data do not include the isotropic extragalctic background (EB), a power-law fit to the isotropic emission model for energies smaller than 100 GeV, were CR contamination is assumed to be small, has been added to the Fermi data to allow for better comparison to the EGRET data for large Galactic latitudes. The same EB model has been used in the aPM prediction. Despite the softer Fermi spectrum the aPM prediction still reveals a slight excess above 1-2 GeV in those regions of the sky where the Galactic emission dominates. More importantly, the overall normalization of the model prediction appears to be too low by a factor ~ 1.5. In the following we will discuss possibilities to improve this situation. Having discussed the EGRET GeV excess in terms of DMA, we will stay in this context and first compare the DM signal derived from EGRET to the Fermi data. In a second approach we will drop the contribution of a possible signal from DMA and discuss the Fermi-LAT data in the context of a purely astrophysical explanation, using constraints on electrons and protons from ATIC and Fermi. In Sections 5.3 and 5.4 we will

[3]In the EGRET data used in the comparison to the aPM previously, the contribution from point sources, such as pulsars, have been subtracted in order to get the diffuse emission. Insofar the EGRET data points shown also constitute a model of the diffuse emission.

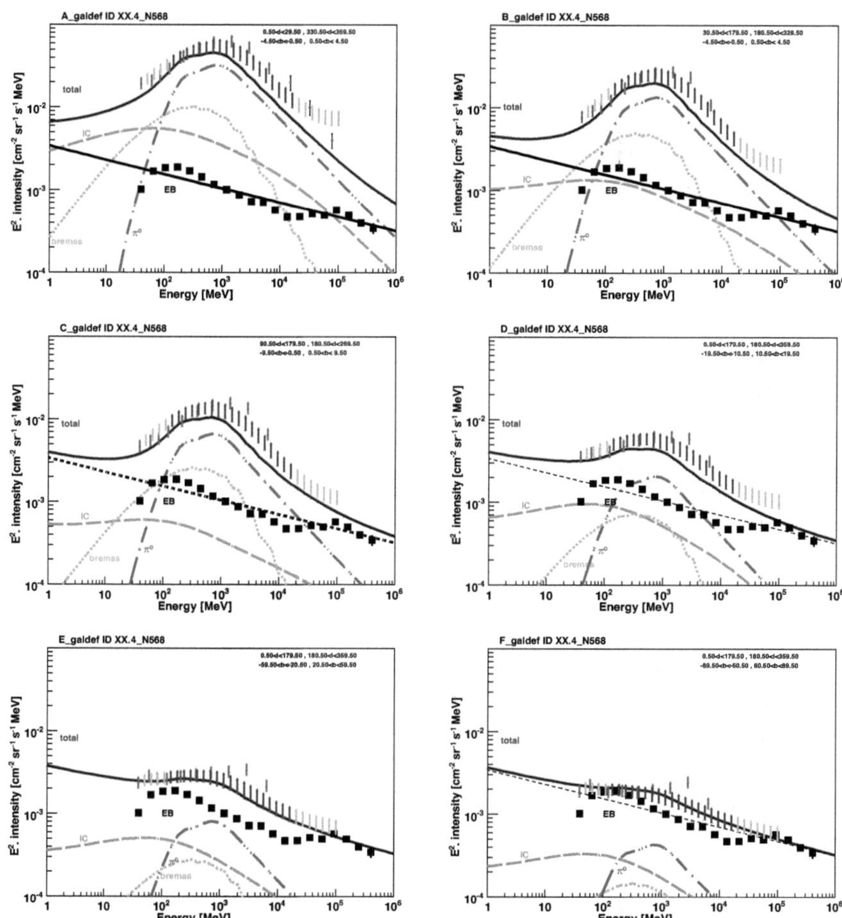

Figure 5.6: The aPM diffuse γ-ray prediction for the six sky regions compared to the EGRET and preliminary Fermi-LAT data. The Fermi point to point errors are assumed to be 15%, similar to EGRET. The red data points are the EGRET data, the blue data points are the Fermi data in the energy range we consider here (120 MeV to 20 GeV, see text for details) and the grey points are the Fermi data outside this range, which are shown for completeness.

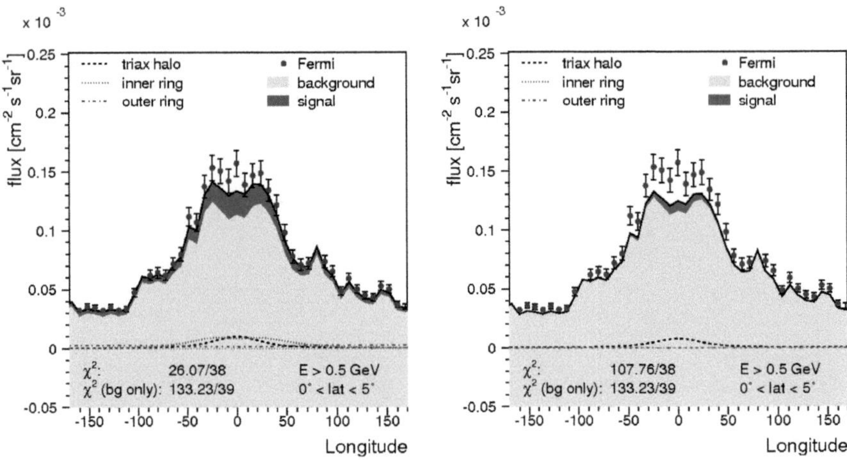

Figure 5.7: Longitude profile for a fit to the preliminary Fermi-LAT data for a halo profile with (**left**) and without (**right**) rings: 0° ≤b< 5°.

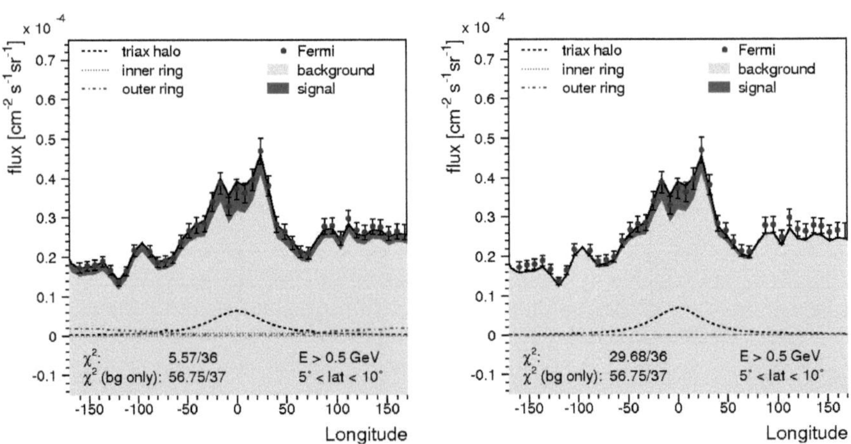

Figure 5.8: Longitude profile for a fit to the preliminary Fermi-LAT data for a halo profile with (**left**) and without (**right**) rings: 5° ≤b< 10°.

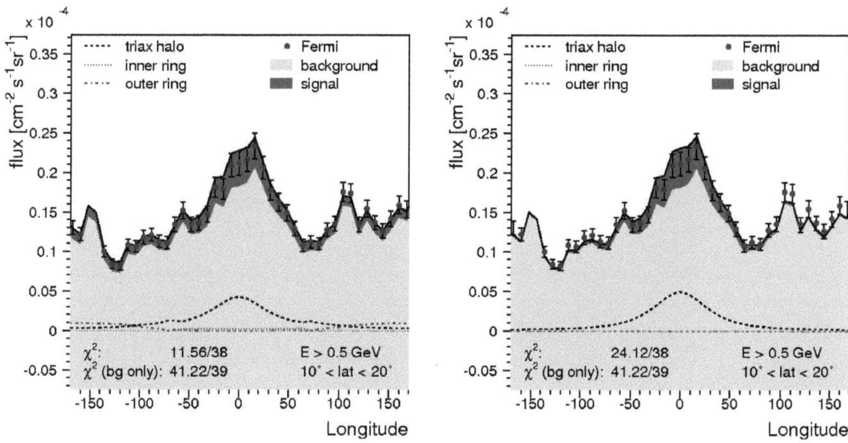

Figure 5.9: Longitude profile for a fit to the preliminary Fermi-LAT data for a halo profile with (**left**) and without (**right**) rings: $10° \leq b < 20°$.

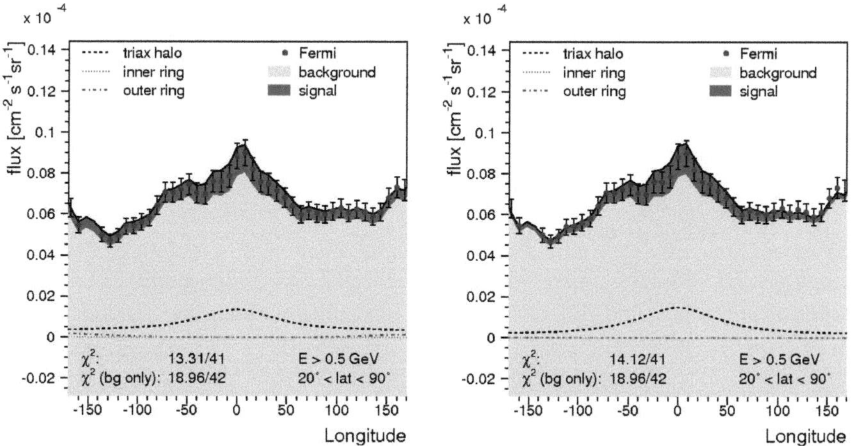

Figure 5.10: Longitude profile for a fit to the preliminary Fermi-LAT data for a halo profile with (**left**) and without (**right**) rings: $20° \leq b < 90°$.

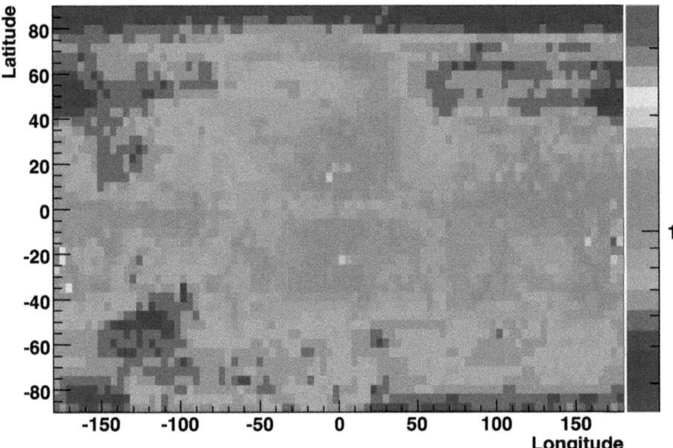

Figure 5.11: Sky map of the background scaling factor for the FERMI-LAT data in a conventional GALPROP model. A 90×45 binning was used. The gas distribution used for the preliminary *gll_iem_v02* model is very similar to the gas distribution in the conventional model. For this reason the scaling factor is close to one in those regions where the emission from the gas dominates. The regions with values above 1 correspond to unsubtracted point sources. The overall scaling factor is around 1 which indicates that the GALPROP predictions are in good agreement with the scaling derived from the fit.

discuss the uncertainties in the absolute normalization of the Galactic γ-ray flux and the local antiproton flux, respectively.

5.2.3 Fermi-LAT and Dark Matter

Since the Fermi-LAT results have been published rather late in the course of this work, a detailed fit of a DM halo profile and DM mass to the Fermi data is yet to be performed. As a first estimate, we can however ask whether or not the Fermi-LAT data are compatible with the halo profile and DM mass as derived from EGRET. Figures 5.7 to 5.10 show the longitude profiles for energies above 0.5 GeV for a pseudo-isothermal profile (PISO) as derived in Sander (2005). The parameters of the profile are given in Appendix C[4]. Only the Fermi-LAT data between 142 MeV and 20 GeV have been used in the fit (see discus-

[4]The fits to the Fermi-LAT data have been performed in close collaboration with Markus Weber, using the halofitter code developed by Christian Sander and Marc Herold (Sander, 2005).

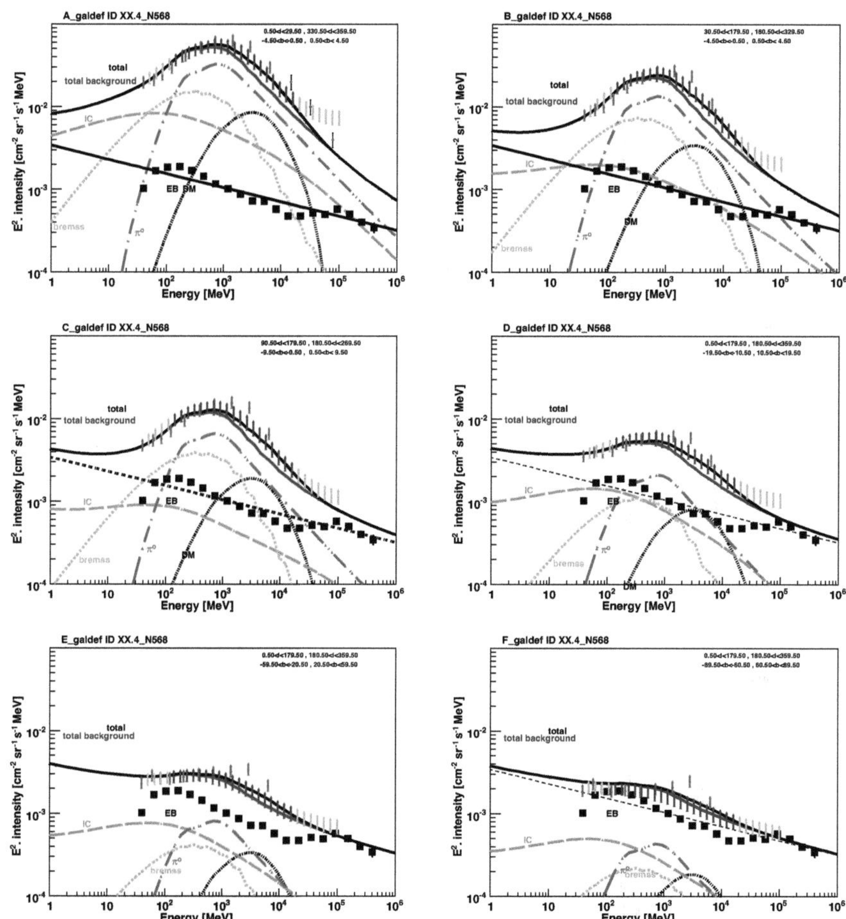

Figure 5.12: The aPM diffuse γ-ray prediction for the six sky regions compared to the EGRET and preliminary Fermi-LAT data. The DMA signal with a boost factor of 10 has been added to the model prediction, the Galactic electron density has been scaled by a factor of 1.5.

sion in Section 5.2.2). The spectral shape of background used for the fits is taken from the conventional GALPROP model (Strong et al., 2004a), an isotropic model with very similar π^0-, bremsstrahlung- and IC-spectra as the aPM. Since here we are only interested in the general question whether or not the Fermi-LAT data are compatible with an EGRET-like halo profile, we can disregard the slight differences between the different background models. The sky has been subdivided into 4050 spatial bins and for each $4^o \times 4^o$ bin the shapes of the background contributions are normalized to the Fermi-LAT data below 0.5 GeV, where any DMA contribution is assumed to be negligible. The resulting scaling factors for the background are shown in Fig. 5.11. Note, that this approach is different from what one usually does in the context of CR modelling: Here we assume that the spectral shape of the background is given by a detailed transport model (the conventional GALPROP model) and determine the absolute normalization for each direction from a fit to the data. This way, features that are present in the data, but are not modelled by the transport model can be included in the background, but at the same time it is not known which astrophysical process might produce these features and, in particular, whether or not this process would also change the spectral shape of the background in the region under consideration. Insofar the background prediction in these fits does not constitute a model prediction, which is based on astrophysical assumptions, but a "data-driven model", where the impact of (unknown) astrophysical effects is incorporated by the background scaling factors. Such a data-driven model, of course, yields a better description of the data than a model based on a limited number of astrophysical assumptions. The background scaling factors derived from the low energy data are then assumed to also hold in the GeV range. This way the precise description of the low energy data can be used to see whether or not an additional contribution from DMA is required by the data. As long as the background scaling factors derived from the fit to the low energy data do not exceed the uncertainties of the gas and CR density distribution in the underlying CR transport model, this approach is valid. From Figs. 5.7 to 5.10 it is clear that the Fermi data are compatible with the halo profile derived from the EGRET data. One notes from the χ^2 values in the figures that the inclusion of the DM halo from the EGRET data significantly improves the fit compared with a fit without DM, called background only in the figures. Especially Fig. 5.7 shows how the inner ring derived from EGRET improves the fit, although further optimization would be required. Since here we are only interested in the question whether or not the preliminary Fermi-LAT data are in principle compatible with the EGRET profile, we refrain from additional fine tuning of the halo profile until the Fermi-LAT data are officially published. In Figs. 5.7 to 5.10 we use error bars of 7%, which corresponds to the point-to-point errors determined for the EGRET data (Sander, 2005). A detailed discussion of the goodness of fit would require to optimize the halo profile for the preliminary Fermi-LAT data and choose the uncorrelated errors in such a way that the total $\chi^2/d.o.f$ adds up to one for the optimized profile.

Having established, that the DM profile derived from the EGRET data is compatible with the Fermi-LAT data we can return to an astrophysical interpretation of the data and compare the CR background prediction of the aPM together with a DMA contribution to the Fermi-LAT data. Figure 5.12 shows the diffuse γ-rays for the six sky regions as defined in Section 4.4.1 for the aPM plus an additional component from DMA. The Galactic electron density has been scaled by a factor 1.5. The halo profile is the profile described in Appendix C, the boost factor [5] derived from the Fermi-LAT data is 10. With the additional component from DMA the spectral shape of the model prediction is greatly improved, however, the total predicted emissivity is still too low. A good description of the data can be obtained with the additional scaling of the electron density. Clearly, with or without an additional signal from DMA the Fermi data either require the Galactic electron density to be significantly larger than what one would expect from the locally measured flux, or the gas density has to be increased. We will discuss effects that might decouple the local CRs and γ-ray production rate in Section 5.3 in more detail.

5.2.4 Fermi-LAT and Astrophysical Explanations

Here we are going to address a purely astrophysical explanation of the Fermi-LAT data on diffuse γ-rays. From the spectral shape in Fig. 5.6 it is clear that the Fermi data still require a somewhat harder spectrum than the aPM currently predicts. A harder γ-ray spectrum can be accomplished by choosing a harder proton and harder electron spectrum, but both these spectra are limited by the locally measured spectra. If one takes into account that the source spectrum of CRs in other parts of the Galaxy might be different from the spectrum of local sources or that the local diffusion coefficient might be not representative for other parts of the Galaxy, one could, in principle, allow for the Galactic proton and electron spectra to be significantly harder than the locally measured proton

[5] Note, that this boost-factor is different from the boost-factors derived from the fit to the Fermi data, because here we keep the background scaling constant for all directions and in addition the local DM density differs slightly (see Appendix C). The boost-factor derived from the fit with spatial variations in the background scaling are 8.61 for the profile with two rings and 10.69 for the profile without rings.

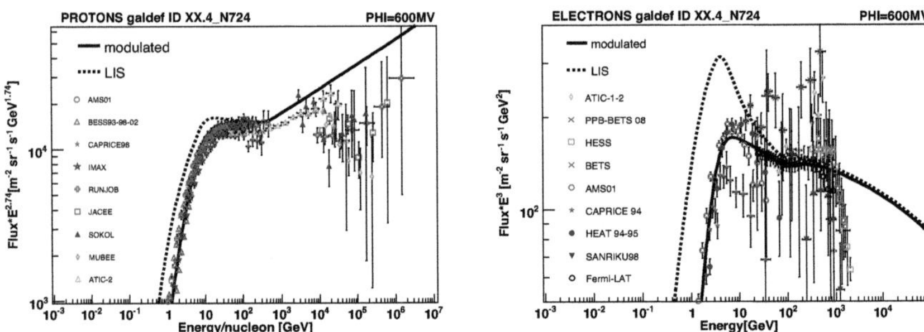

Figure 5.13: Locally measured proton (**left**) and electron (**right**) spectra with the injection indices optimized for ATIC (protons) and Fermi (electrons). The electron flux is normalized to the Fermi data, the proton flux is normalized to BESS.

Table 5.2: Injection spectra of electrons and protons for the Fermi and ATIC data. Two breaks at ρ_1 and ρ_2 are used, $\alpha/\beta_1 - \alpha/\beta_3$ are the correspoonding injection indices.

Protons/nuclei	
$\alpha_1/\alpha_2/\alpha_3$ ρ_1^p/ρ_2^p	1.6/2.41/2.25 9 GV/400 GV
Electrons	
$\beta_1/\beta_2/\beta_3$ ρ_1^e/ρ_2^e	1.6/2.54/2.3 4 GV/100 GV

and electron spectra. Recent results of the Fermi-LAT collaboration strongly favor only slight variations of the spectra (Abdo et al., 2009b). We therefore use the hardest locally measured proton and electron spectra to limit the Galactic spectra. The aim of this study is to determine whether or not the locally measured electron and proton spectra are *in principle* compatible with the Fermi-LAT data on diffuse γ-rays.

The Fermi and ATIC results on electrons and protons

Figure 5.13 shows the locally measured electron and proton spectra. For the protons the hardest spectrum available is the ATIC measurement with a slope of 2.64. However, this data has to be handled with due care: ATIC is a calorimeter experiment and without a charge measurement prone to background from low energetic heavy nuclei. Note that ATIC measures an increase above the index from spectrometer experiments for both the electron and proton/nuclei spectra. Just at the edge of acceptance the spectra then fall sharply.

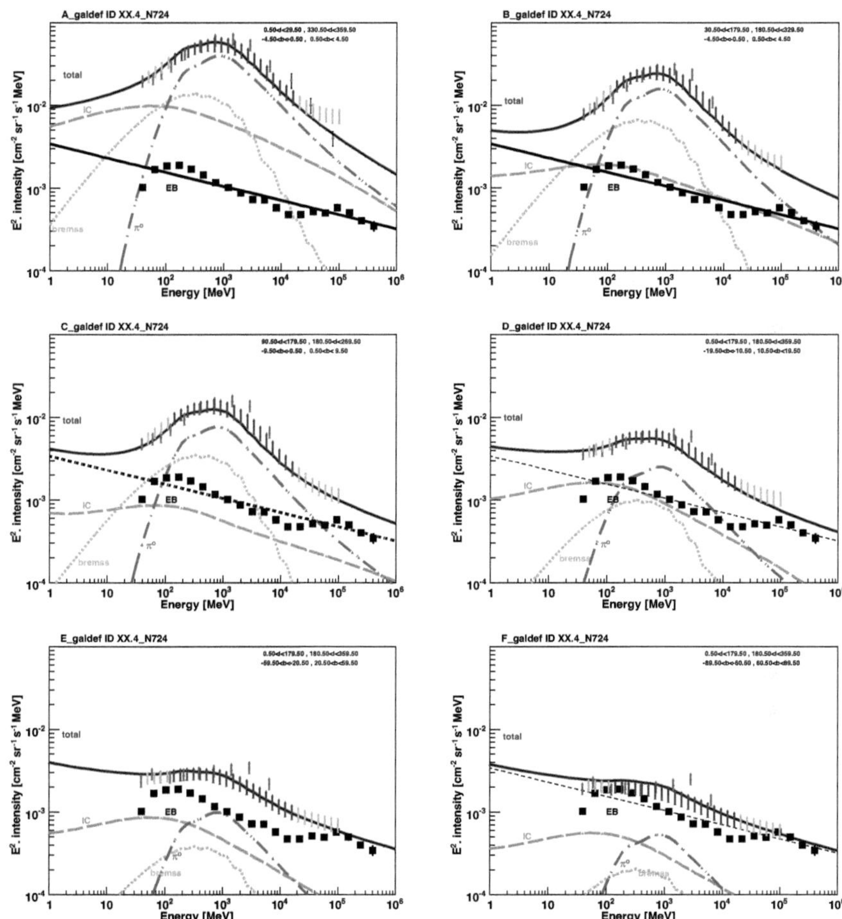

Figure 5.14: Diffuse γ-rays in an aPM with hard electron and proton spectra. The Galactic proton density has been increased by a factor 1.2 compared to the normalization of the local proton flux, the Galactic electron density has been increased by a factor 1.5 compared to the local electron flux.

Since here we are interested in the hardest possible proton spectrum we will put these doubts aside for now and assume that the local proton spectrum has a slope compatible with the ATIC measurement. Since the normalization between different experiments can be off, we apply the same arguments to the absolute value of the local proton flux and normalize the model prediction to the high energy BESS data. This way we make sure to get the highest and hardest π^0-prediction, which can still be considered compatible with the local proton measurements.
For electrons we use the Fermi-LAT measurement since it covers the largest energy range with the smallest errors. It also allows us to keep the correlation between the diffuse γ-ray data and the electrons. The usage of the ATIC data might appear more reasonable since it would correlate proton and electron fluxes, but the contribution from high-energy electrons to the γ-rays is insignificant compared to protons. Also note that for γ-rays below 1 GeV the normalization of the local electron spectrum is most important and not so much the spectral shape. Given the large ATIC error bars the normalization of the Fermi and ATIC data is compatible. Figure 5.13 shows the predicted proton and electron flux in an aPM with the injection indices optimized for the respective data. The injection indices used here are given in Table 5.2. The change in injection index does not influence the B/C and $^{10}Be/^{9}Be$ ratios since here the spectral dependence cancels.
Figure 5.14 shows the diffuse γ-rays for the six sky regions. The spectral shape of the model prediction is improved compared to model prediction for the softer proton and electron spectra shown in Fig. 5.6. The overall deficiency remains, but after scaling the Galactic proton density by a factor 1.2 and the Galactic electron density by a factor 1.5, as was done in Fig. 5.6, we find the γ-rays well-described within the assumed errors of 15%.

5.3 On the Link between Local Charged CRs and Diffuse γ-Rays

In the previous sections we have discussed the Fermi-LAT data in the context of DMA and in the context of a purely astrophysical explanation. In both cases the normalization of local electrons and protons and the diffuse γ-rays do not agree. Possible reasons for this discrepancy can be variations in the CR density distribution (such that the local CR density is lower than the Galactic average) or the gas distribution (again the local gas density would have to be lower than the averaged density). There also might be an additional γ-ray contribution from large distances, i.e. the region beyond the mathematical

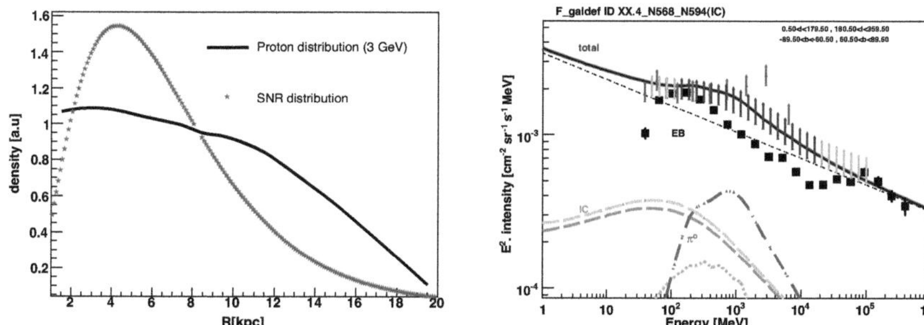

Figure 5.15: **Left:** Radial proton distribution for 3 GeV protons for a model with zero source strength between $R = 8.2$ kpc and $R = 8.4$ kpc (black full line). Also shown in the normalized source distribution (red marker) with the gap between 8.2 kpc and 8.4 kpc. **Right:** Diffuse γ-rays in the aPM (z_h=7.5 kpc) for region F. The grey line shows the IC contribution for a run with z_h=30 kpc.

boundary of our numerical solution. The most obvious solution of course would be a locally decreased source strength. Naively one would expect that this way the local CR density could be decreased by a factor $\sim$1.5, so that the normalization of local CRs and diffuse γ-rays would be in agreement. In Section 4.3.2 we have shown that the collection distance for CRs in an aPM and an isotropic model are similar. Furthermore, with more than 50% of the protons between 10 and 10^3 GeV originating from distances larger than 1 kpc we find it hard to believe that gradients in the source distribution or the transport parameters, which would occur on scales of parsecs, would not be washed out. Figure 5.15 shows the radial proton distribution for a model with zero source strength between $R = 8.2$ kpc and $R = 8.4$ kpc (i.e. the Local Bubble region, see Section 5.3.3). Clearly, CR transport is efficient enough to remedy such small scale variations in source strength. In the following we will first estimate the maximum additional γ-ray flux from an extended halo and find that this contribution is negligible (Section 5.3.1). We will then estimate the amount of additional gas required by the Fermi-LAT data in Section 5.3.2 and identify structures in the ISM that would reduce the local gas density in sections 5.3.3 and 5.3.4.

5.3.1 The γ-ray contribution from the halo

In order to solve the obvious discrepancy between the local normalization of charged CRs and the Galactic γ-ray emissivity it has been suggested that a large Galactic halo would give rise to an additional IC signal. Since the scale height of the gas distribution is around 250 pc, only little bremsstrahlung or π^0-decay emission is expected from above the Galactic disk. As we have discussed in Section 3.5.3 isotropic diffusion models are very sensitive to the size of the transport box, since this basically constitutes the diffusion-convection boundary. The local B/C and $^{10}Be/^{9}Be$ ratio are best reproduced for a halo height of 4-4.5 kpc, thus the assumption that the IC contribution from distances above the halo boundary might increase the diffuse γ-ray flux stands to reason. In the aPM it is possible to increase the halo to arbitrary sizes, as we have discussed in Section 4.3.1. We have also mentioned that the additional IC contribution from regions above 7.5 kpc is at the percent level. Above $\sim$ 10 kpc the contribution from starlight and dust emission to the ISRF drops below the CMB radiation, which is spatially constant. Consequently, from this height on the IC emissivity drops proportional to the electron density. Figure 5.15 shows the IC contribution for a run with z_h =7.5 kpc (grey line). The increase in flux from IC is below 10% and the corresponding increase in total γ-radiation is negligible. Note, that the increase in IC emission from region F sets an upper limit on the additional IC emission for the other regions, since the relative contribution from the halo is maximal for this region. Insofar an increased IC contribution from a large halo can be excluded as the reason for the discrepancy between the normalization of local CR fluxes and diffuse γ-rays. Also one should not forget that the deficiency in γ-ray flux is independent of direction. An increased halo would, if at all, contribute at high and intermediate latitudes, but not in the Galactic center region.

5.3.2 Untraced Gas Components

Untraced gas components have been suggested as a possible source of additional γ-radiation. "Untraced" gas components could be e.g. H_2 untraced by CO. Assuming that the "untraced" gas component does not follow a specific distribution, but just increases the overall Galactic gas density by a factor χ_{gas} we find that values around $\chi_{gas} \approx 1.5$ are necessary to explain the observed γ-radiation. Figure 5.16 shows the diffuse γ-rays for an aPM with the Galactic gas density globally increased by 50%. Of course, an increase in the gas density also affects the local secondaries, simply because the number of CR interactions with the ISM is increased by the same factor. The left side of Fig. 5.17 shows the local B/C ratio for this model, the right side of the same figure shows $^{10}Be/^{9}Be$, which is also increased because the fraction of locally produced beryllium increases and thus CRs "appear younger". Usually, one would now try to remove the excess secondaries by an increase in the diffusion

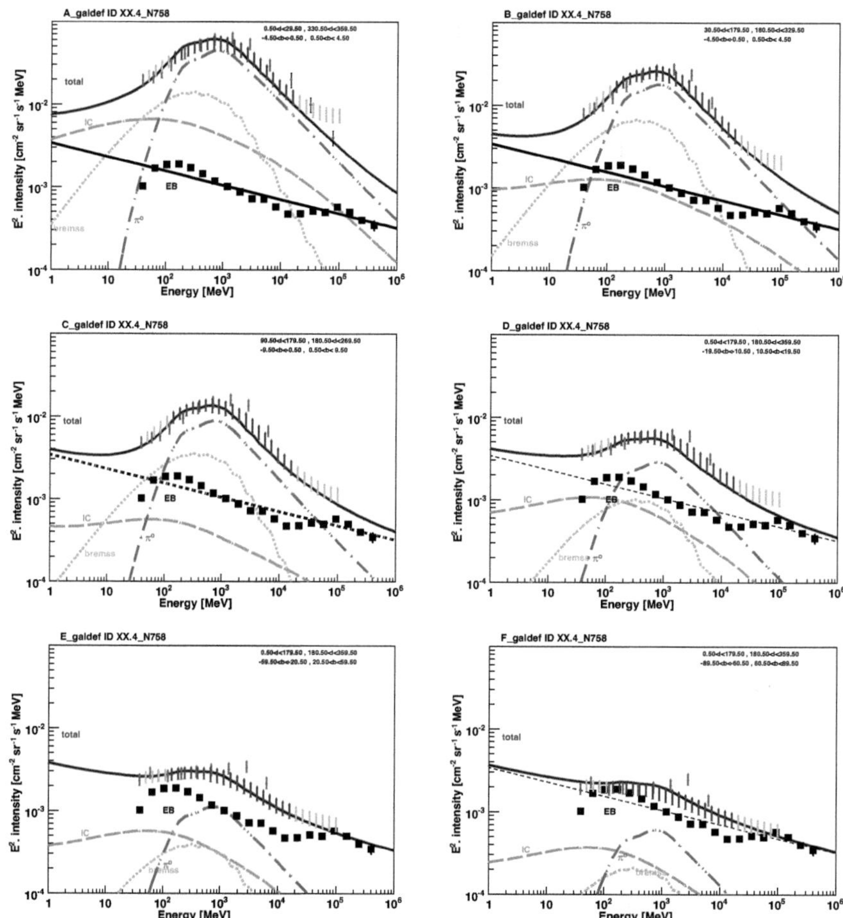

Figure 5.16: The aPM diffuse γ-ray prediction for a model with the Galactic gas density increased by 50%.

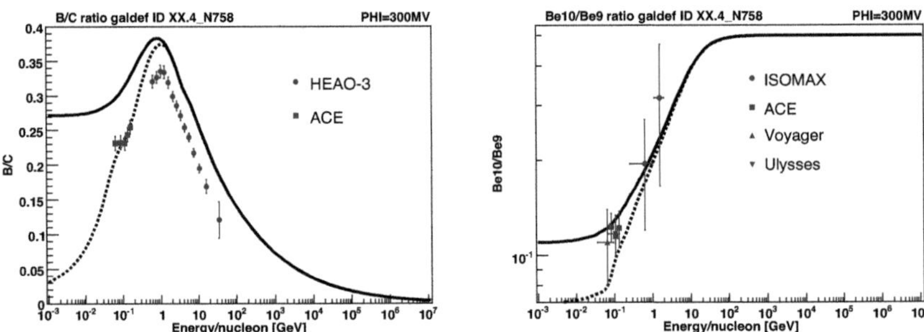

Figure 5.17: The local B/C and $^{10}Be/^{9}Be$ ratio for a model with the Galactic gas density increased by 50%.

coefficient or a decrease of the halo height, but both these adjustments would also decrease the γ-ray flux. This is clear because both modifications would lead to a globally larger CR flux towards the boundary. From here it is obvious, that any process that reduces the local secondary flux has to be a local process insofar that it does not influence the diffuse γ-ray production rate. In the following we will discuss two features of our local environment, the Local Bubble and the local interarm region, that might reduce the local secondary production rate without changing the averaged γ-ray production rate significantly.

Of course, the assumption of an untraced gas component which makes up about one third of the total Galactic gas mass is unlikely. Clearly there must be other effects which also contribute to the increased γ-ray production rate (or the decreased local secondary production rate). Since we do not know these effects, we simply estimate the maximum impact of the known structures of our Galaxy to see whether or not these structures *in principle* could explain the observed γ-radiation.

5.3.3 The Local Bubble

It is well known that our Sun resides in the Local Bubble (LB), a low density region of space extending about 200-300 pc into the Galactic plane and 600 pc perpendicular to it. The LB probably began to form more than 10 Myr ago and is believed to be the result of more than 14 local supernova explosions since then (Fuchs et al., 2006). This region is associated with Gould's Belt (Frisch, 2009), a region of young stars bounding the void.

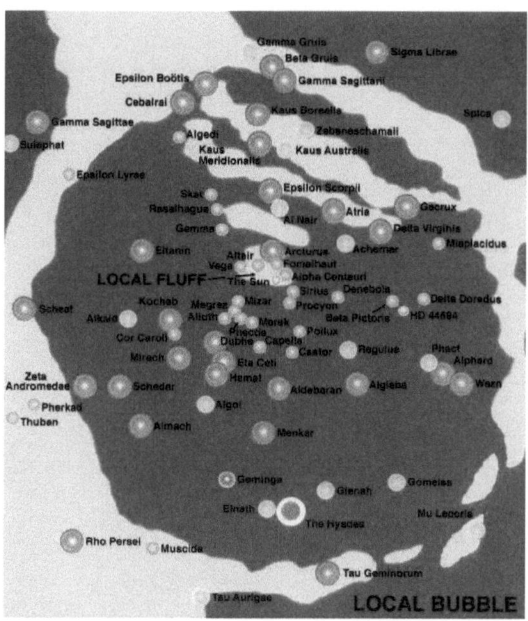

Figure 5.18: **Left:** Local Bubble and Local Fluff. The blue regions depict regions with decreased gas density relative to the yellow regions. The image is available under the terms of CC-by-sa, credit: N. Henbest/H.Couper.

The density inside the Local Bubble is about 0.05 $\frac{atoms}{cm^3}$, which is approximately a tenth of the average density of the ISM. The Sun entered the LB more than 5 million years ago and currently moves through the Local Fluff (LF) complex of interstellar clouds, a slightly denser region within about 35 pc around the Sun. The density of the LF is with 0.26 $\frac{atoms}{cm^3}$ about five times the density of the LB. The LF coincides with the region where the LB and the Loop I superbubble overlap (Frisch, 2009), thus the ISM surrounding the Sun is part of the shell of a superbubble expanding into the low density interior of the Local Bubble. The LB and LF are illustrated in Fig. 5.18

We have implemented the LB and LF in the GALPROP code in the form of two user-defined regions of arbitrary size and arbitrary spatial resolution, in which the gas density and all transport parameters can be reduced or increased by a certain factor f_{LB}^{par} or f_{LF}^{par}, where *par* stands for the transport parameters, the gas density or the source strength.

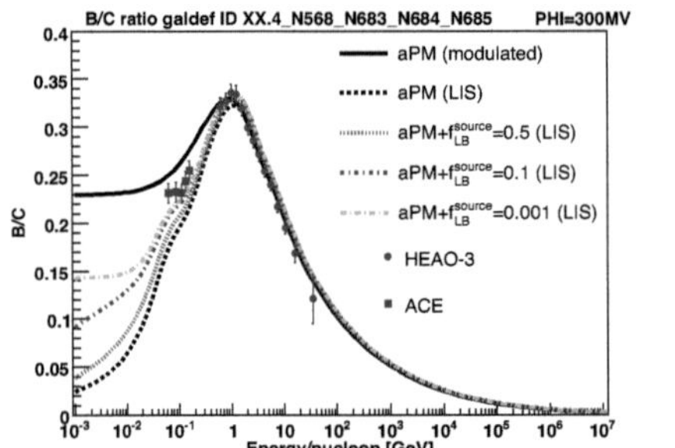

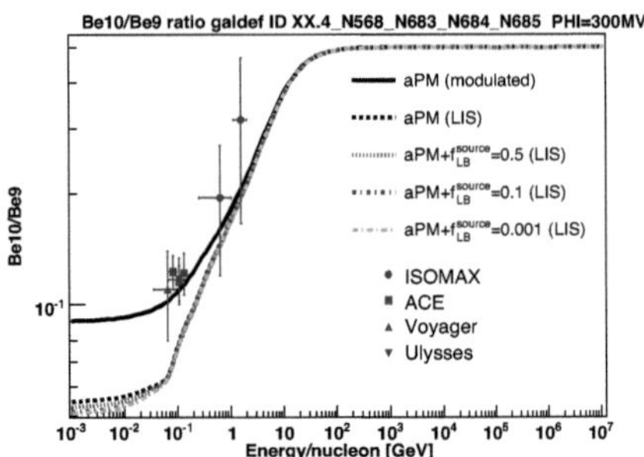

Figure 5.19: Locally decreased source strength: B/C (**left**) and $^{10}Be/^{9}Be$ (**right**) ratio for models with $f_{LB}^{source} = 0.5 - 0.001$ compared the aPM. The size of the LB region is $h_{LB} = 0.2$ kpc and $d_{LB} = 0.2$pc. For LB models only the LIS spectra are shown, since the uncertainties in solar modulation could absorb the effect of the reduced source strength.

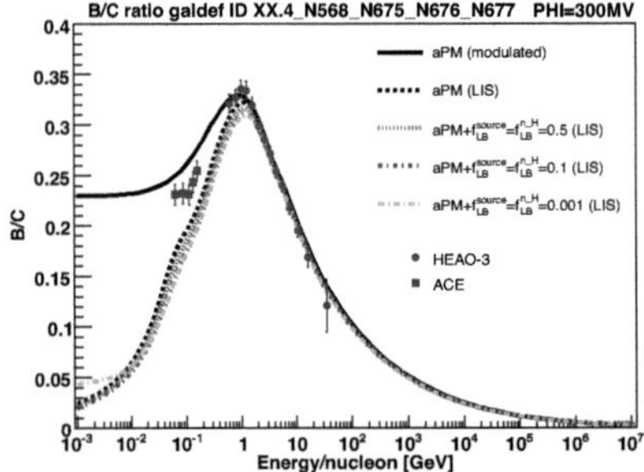

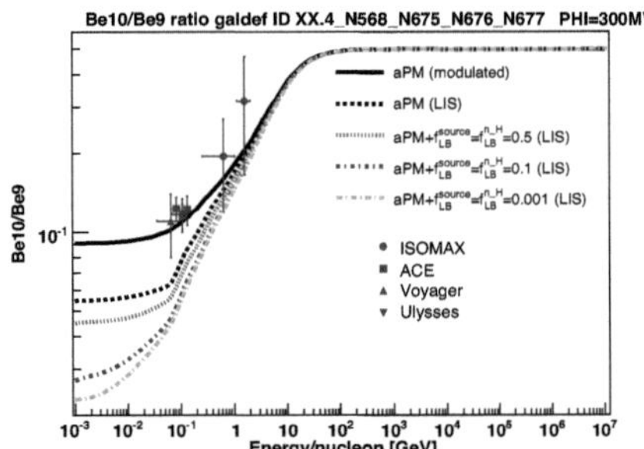

Figure 5.20: Simultaneous decrease in local gas density and source strength: B/C (**left**) and $^{10}Be/^{9}Be$ (**right**) ratio for models with $f_{LB}^{source} = f_{LB}^{n_H} = 0.5 - 0.001$ compared the aPM. The size of the LB region is $h_{LB} = 0.2$ kpc and $d_{LB} = 0.2$pc. For LB models only the LIS spectra are shown, since the uncertainties in solar modulation could absorb the effect of the reduced source strength.

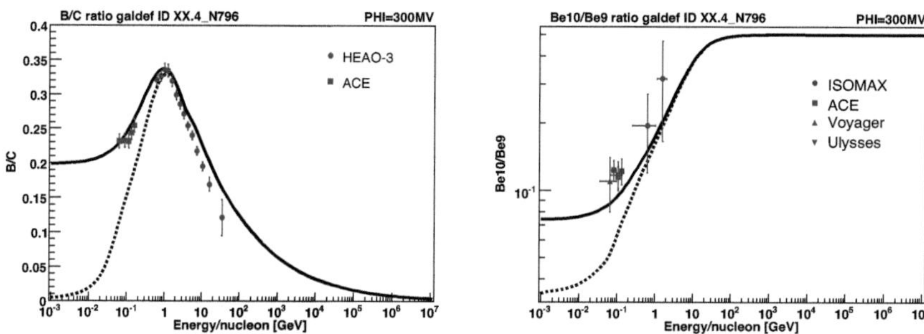

Figure 5.21: B/C (**left**) and $^{10}Be/^{9}Be$ (**right**) ratio for a model with Galactic gas density increased by a factor 1.5, as required by the preliminary Fermi-LAT data and a LB with h_{LB}=0.1 kpc, d_{LB}=0.4 kpc and $f_{LB}^{n_H} = 0.1$.

Note, that in a 2D model as it is used here a LB or LF region actually corresponds to a ring-like structure. This does impose an error upon our estimates: with no gradient in the source distribution, the gas distribution, the diffusion coefficient and all relevant fields, the problem is symmetric and no resulting flux can occur (although, of course, single particles will change places). If gradients in any of the above are present, the propagated CR distribution will adapt to these gradients at the border of the LB or LF region and thus slightly modify the result. Given the high memory requirements of a 2D implementation of the aPM and the additional memory required for a local high resolution region, a 3D implementation is beyond the scope of the available computing resources[6]. Since we do not intend to model all aspects of our local environment (which are anyway unknown on a level of CR transport parameters), we can airily reduce the level of accuracy and stay in cylindrical symmetry.

Naturally, secondary to primary ratios are especially sensitive to local changes in the gas density. With lower local gas densities the amount of local secondaries will decrease, which will reduce both, the B/C ratio and the $^{10}Be/^{9}Be$ ratio. The degeneracy between the different parameters is large, e.g. a decrease in gas density can be counterweighted by a decrease in diffusion coefficient and the appropriate change in v_α, and so a detailed modelling of the Sun's local environment would be without physical meaning. A further

[6]We have currently one machine with 32 GB memory, which can be used for extensive scans.

complication arises from the fact that the source strength inside the LB might drop to very small values, since there are no known sources for CRs inside this region. We have already seen that even a reduction in source strength to zero has almost no effect on the propagated proton distribution, at least at relevant energies. For the local B/C ratio a decrease in local source strength will increase the B/C ratio especially at low energies, because the amount of low-energy C that can reach the Earth from outside the LB will be reduced (see the left side of Fig. 5.19). For $^{10}Be/^{9}Be$ only a negligible effect is expected, since the secondary production rate is not changed (see the right side of Fig. 5.19). This changes not only the source distribution, but also the gas distribution is reduced. Figure 5.20 shows $^{10}Be/^{9}Be$ for $f_{LB}^{n_H} = f_{LB}^{source}$, the reduction in $^{10}Be/^{9}Be$ is due to the fact, that inside the LB ^{10}Be can decay, but only little beryllium can be produced. For B/C on the other hand a simultaneous decrease in gas density and source strength does not affect the ratio.

Since here we are interested in a reduction of the amount of secondaries, we just examine the impact of a reduced gas density in a region with vertical height h_{LB} and radial extension d_{lb} upon the local B/C ratio. From Fig. 5.16 we know that the local B/C ratio has to be reduced by a factor of ~ 1.15. Figure 5.21 shows the resulting B/C (left) and $^{10}Be/^{9}Be$ (right) ratio for a global gas density increased by a factor 1.5 and a local gas density decreased by a factor $f_{LB}^{n_H}$=0.1 in a region with h_{LB}=0.2 kpc and d_{lb}=0.4 kpc, which means that we choose a slightly larger diameter for our LB than the usually quoted 200-300 pc. The overall amount of secondaries is well reproduced, only the spectral shape of the B/C ratio requires some minor fine tuning above 1 GeV. The local age of CRs is somewhat too large as can be seen from the rather low $^{10}Be/^{9}Be$ ratio. The reason for this is that inside the LB ^{10}Be can decay, but too few Be is produced. Additional fine-tuning of the model can be done via f_{LB}^{D} and $f_{LB}^{v_a}$, however, for this level of detail a 3D implementation would be required.

Nevertheless, with this simple estimate we have demonstrated that the difference in normalization of the diffuse γ-rays and the local CR density can be understood in the context of variations in the gas density. Although the reduction in local gas density is compatible with what is expected from the LB, the increase of Galactic gas density by 50% has currently no immediate physical motivation. In the next section we will discuss the spiral structure of the Milky Way, which is likely to help in this matter.

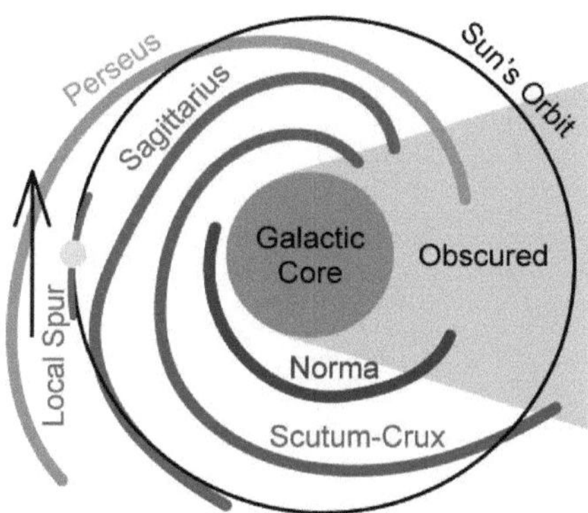

Figure 5.22: Observed spiral structure of the Milky Way galaxy following Taylor & Cordes (1993). The Sun is indicated by the yellow spot, the arrow indicates the direction of the solar system's motion relative to the spiral arms. The plotted orbit of the sun does not account for perturbations caused by interactions with the spiral arms. Source: Wikipedia.

5.3.4 The Spiral Structure of the Milky Way

Here we will address a feature of our Milky Way that might reduce the effective local gas density and increase the effective Galactic gas density and in this way help to reduce the amount of "untraced" gas components. The spiral structure of the Milky Way, i.e. the fact that the gas and source distribution is aligned along the spiral arms, is generally neglected in CR transport models. It does enter the calculation of the diffuse γ-ray emission in GALPROP by virtue of the normalization to the HI and CO astronomical surveys, but the gas distribution used for the propagation of CRs is rather smooth, as discussed in Section 3.3.1. In reality the gas and the SN distribution are closely linked to the spiral structure of the Milky Way, which is obvious since the spiral arms consist of young stars and molecular gas. The Sun happens to reside just between two spiral arms, the Perseus arm and the Sagittarius arm, as can be seen from Fig. 5.22. This means that the effective gas density is probably lower than the smoothed gas density used in GALPROP. The exact alignment of the Milky Ways spiral arms is still under discussion (see Hou et al. (2009) for an overview). However, there seems to be convergence towards a model with four spiral arms (Vallée, 2005). To estimate their impact on CR transport the spiral arms can be simply modelled by a set of Gaussian rings in the Galactic plane with a width of 2 kpc.

We model the four spiral arms by four rings with Galactocentric distances

- **Cygnus:** 13 kpc
- **Perseus:** 10 kpc
- **Sagittarius:** 7.5 kpc
- **Crux+Norma:** 4.5 kpc

which approximately correspond to the Galactocentric radii of the respective arms in the direction of the Sun. The gas density in the spiral arms can be assumed to be a factor 10 higher than in the interarm region. For GALPROP this means that the local gas density is greatly reduced in a region much larger than the LB, since the Sun is positioned at the tails of Perseus and Sagittarius. It is obvious that redistributing the Galactic gas content in the form of spiral arms can either increase the flux of diffuse γ-rays or decrease the local amount of secondaries: In the latter case the local reduction of gas density will decrease the B/C ratio. To balance the secondary production rate the diffusion coefficient can be reduced, which means that CRs stay for a longer time in the Galaxy. This way more CRs will occupy the gaseous disk in the steady-state case and consequently the γ-ray emissivity, which is given by the CR (proton or electron) density times the gas distribution or the respective field, will be increased.
Although the above idea is quite promising, there are other effects which have to be taken into account:

- Since the molecular phase of the ISM is accompanied by its own small-scale magnetic field, variations in the H_2 distribution are expected to be accompanied by a corresponding variation in the diffusion coefficient and Alfvén velocity. This is currently not modelled in our code.
- If the source strength is modulated on a kpc scale, the same modulation is expected in the wind strength, since in regions between the spiral arms, i.e. regions with low source density, the CR pressure might not suffice to launch a wind. In addition a resulting radial wind component V_{RR} towards the outer Galaxy would be expected in the regions with low source strength. Both these effects are currently not modelled in our code.

Similar to the case of the LB, at this level of detail a full 3D implementation would be required to take into account **all** the effects expected from the respective features. The extended GALPROP version developed for the aPM is the first transport code capable of examining these structures. It will be a valuable tool in the detailed examination of small scale structure in the future.

5.3.5 Some concluding Notes on the Link between diffuse γ-rays and local charged Cosmic Rays

We have seen that diffuse γ-rays require a Galactic gas density increased by a factor of 1.5 in order to be compatible with the Fermi-LAT data. Such an increase in gas density would increase the amount of local secondaries to an unacceptable level. A global increase in diffusion would decrease the number of produced secondaries, but at the same time the number of produced γs will be decreased. As a possible solution we discussed variations in the local gas density which would decrease the *locally* produced amount of secondaries, while keeping the *global* γ-ray production rate almost constant.
For the gas distribution errors of 10-20% are generally accepted. Insofar the assumption of untraced gas components which add up to 50% of the traced gas components appears rather extreme, but remember, that here we were only interested in a rough estimate of the size of this effect. In particular we neglected possible variations in the diffusion coefficient in the disk. A more detailed modelling of the geometry of the Milky Way, especially the spiral arm structure, is expected to help in this context. Furthermore, a radial dependence of the diffusion coefficients or an additional anisotropy in diffusion, which have been neglected so far, might improve the diffuse γ-ray prediction. Unfortunately, almost nothing is known about the spatial dependence of the diffusion tensor, so that a large number equivalent scenarios have to be tested. For example, a smaller radial diffusion coefficient D_{RR} in the plane would increase the CR density and consequently the γ-ray emission from the plane. Yet, at the same time the gradient of the diffuse γ-ray emission would become harder and the soft-γ-ray gradient-problem would reappear. In order to keep the γ-ray gradient flat one would have to resort to a rather fine-tuned scenario of diffusion vertical to the plane and convection. Alternatively, one could imagine that the vertical diffusion coefficient D_{zz} is smaller for Galactocentric distances other than $R_{\odot}$. This assumption is motivated by the fact that the Sun resides in the local interarm region (modelled by D_{zz}^{loc}) while D_{zz}^{av} models the averaged diffusion coefficient of spiral arms and interarm regions). A smaller D_{zz}^{av} would lead to a smaller CR flux towards the halo boundary and consequently to a higher CR density in the Galactic plane (and therefore more γ-ray emission), while the local age of CRs and their grammage would still be given by D_{zz}^{loc}.

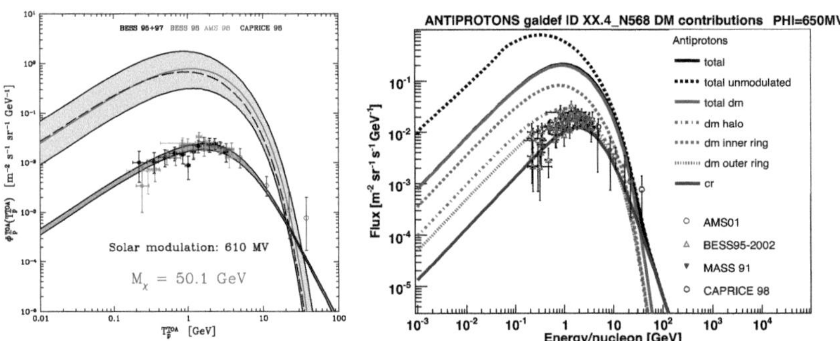

Figure 5.23: **Left** Local antiproton prediction from DarkSUSY. Taken from Bergstrom et al. (2006). **Right** Local antiproton flux in an aPM using a boost factor of 27. The contribution from CRs is very similar to the predictions of the isotropic models. The DM contributions from halo and rings are shown separately.

Many more scenarios are possible and none of them can be prioritized by first principle. Here we have developed a GALPROP version, which is capable to account for spatial variations in basically all transport parameters and therefore allows us to test these scenarios in future studies. On the side of the data a great improvement is expected from the AMS-02 detector, which will be launched mid 2010. AMS-02 will provide us with data on all components of charged CRs and γ-rays. Having data on B/C, protons, electrons, positrons and antiprotons from the same detector will greatly improve our understanding of the transport processes in our local environment and together with the diffuse γ-rays we will be able to better constrain the global transport parameters and thus the number of possible scenarios.

5.4 Constraints from Antiprotons

Having discussed the problems and possible solutions of diffuse γ-rays, we now turn to the local antiproton flux, which constitutes a problem in both, purely astrophysical models and models which invoke an additional contribution from DMA. We have already seen that the antiproton flux from CRs is too low. In fact, most GALPROP models usually predict local antiproton fluxes from CRs about 40% too low and the aPM is no exception as we have seen in Section 4.2. There are simple analytical models like DarkSUSY, that are

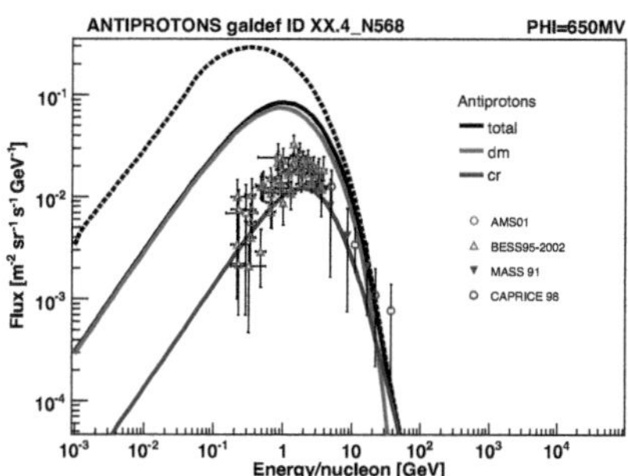

Figure 5.24: Local antiproton flux in an aPM for a EGRET compatible halo profile and a boost-factor of 10.

able to describe the local antiproton flux well without an additional antiproton component from DMA. These models are strongly simplified analytical or semi-analytical solutions of the diffusion equation, close to the leaky-box approach, and feature only few details of the actual geometry of the Milky Way, like the gas distribution and the magnetic field. In this situation an additional antiproton contribution from DMA might be considered a welcome addition, but the contribution from DMA generally overshoots the data as we will see in the following. The left side of Fig. 5.23 shows the antiproton flux from DMA for a 60 GeV neutralino and a halo profile as derived from the EGRET data. The fact that the antiprotons overshoot the data by a factor of $\sim$ 37 has been considered a major argument against the DMA interpretation of the EGRET excess by Bergstrom et al. (2006). The authors used the simple analytical solution of the transport equation in DarkSUSY for their prediction of the local antiproton flux. One could now expect that a model with strong wind velocities would significantly reduce the local antiproton flux from DMA, since antiprotons produced in the large DM halo are unlikely to reach Earth. We have implemented additional CR components for antiprotons, protons, electrons and positrons from DMA in GALPROP. The yield per annihilation of the respective particle is taken from DarkSUSY and gives the injection spectrum, the DM halo profile is chosen to agree with the EGRET excess and gives the source distribution for these particles. The subsequent propagation is identical to the propagation of conventional CR antiprotons, protons, electrons and positrons. Charged species from DMA and charged species from CRs are propagated as separate densities, which is valid, since the CR transport equation is linear in CR density and so the gradient in the CR proton source distribution does not influence the propagation of the DMA protons and vice versa. Since the source strength for CRs from DMA is known (i.e. given by the DM density and boost-factor derived γ-ray signal) no normalization is applied at the end of the propagation. The right side of Fig. 5.23 shows the local antiproton flux for an aPM with the halo profile as derived from the EGRET DMA interpretation and a boost-factor of 27. The antiproton flux from nuclear interactions tends to be on the low side compared to the data, while the additional component from DMA exceeds the data by a factor of ~ 8. Also shown are the separate contributions from the different halo components to the antiproton flux. From Fig. 5.23 one can see, that the inner ring at 4 kpc is the major contributor to the local antiproton flux. Compared to the simple DarkSUSY prediction in the aPM the local flux from DM antiprotons is slightly reduced due to the slightly decreased diffusion coefficient, but the model predicts still a factor of ~ 8 too many antiprotons.

Figure 5.24 shows the antiproton flux at the position of the Earth for a boost-factor of 10 as expected from the Fermi-LAT data on diffuse γ-rays (see Fig. 5.12). Since the boost-factor used here is independent of position, the reduction in this parameter of course directly enters the local antiproton density. The model prediction is therefore only a factor ~ 3.2 above the data.

This means that for each photon from DMA around 1 (for a boost-factor of 10) or around 3 (for a boost factor of 27) antiprotons from DMA arrive at Earth. Given the fact that only around 0.01 antiprotons per photon are produced during the annihilation, this is a rather surprising result. However, it is understandable if one considers the transport effects that affect the antiprotons, but not the γ-rays: Unlike the γ-rays which follow straight lines, the antiprotons immediately begin to scatter on the magnetic turbulences generated by the CR plasma. Since the source strength of CRs from DMA is fixed by the yield per annihilation and the DM annihilation rate, i.e. the γ-ray signal, the absolute flux of these CRs depends on the transport parameters and the source distribution. Note that this is different from "conventional" CRs for which the source strength is not well constrained and which are normalized to the local CR fluxes after propagation. More specifically the local antiproton density depends on the drift velocity of CRs toward the boundary, which, given the fact that annihilation in the ISM in negligible for antiprotons, is the only loss mechanism for this species. Large drift velocities will lead to fast particle escape and, due to the fixed production rate, to low local antiproton densities in the steady state. However, the transport parameters, and thus the drift velocity, is fixed by the measurements of secondary to primary ratios and radioactive isotopes as we have seen in Chapter 3. The local measurements of B/C and $^{10}Be/^{9}Be$ require CRs to spend a certain amount of time in the thin halo and in the gaseous disk. A low antiproton flux from DMA would require fast CR escape from the Galaxy, but for a given halo height (or diffusion convection boundary) this is tightly constrained by the ratio of $^{10}Be/^{9}Be$ (see the comment on unstable isotopes in Appendix A.7). Likewise an increase in halo height, which would allow for large drift velocities, is forbidden by the local B/C and $^{10}Be/^{9}Be$ ratio, which constraints the relative amount of time spend in the halo and in the disk. Thus, for a given gas distribution, these two measurements fix the drift velocity, leaving only small uncertainties in the local antiproton flux from DMA. One of these uncertainties originates from the difference in source distributions. For CR antiprotons this is the gas distribution times the proton distribution, for DM antiprotons, this is the squared DM distribution. For the DM profile derived from the EGRET data the sun resides in a local minimum (see Fig. 5.2). From the right side of Fig. 5.23 it is clear that most of the local antiprotons from DMA originate from the inner ring at 4 kpc. This means that it would be preferable if the collection distance for antiprotons (and consequently also for all other nuclei) would be as small as possible. That way, the inner ring would contribute less to the local antiproton flux. We will discuss these uncertainties in the antiproton flux from DMA in Section 5.4.1.

5.4.1 Disentangling B/C and Antiprotons from Dark Matter Annihilation

Anisotropic Diffusion

A reduction in collection distance requires slower CR transport. In radial direction diffusion is the only transport mechanism and consequently a reduction in diffusion coefficient is required. If the convection velocity is kept compatible with the ROSAT observations any change in diffusion coefficient will also change z_c, as well as B/C and $^{10}Be/^{9}Be$. The additional freedom from local variations of the gas density is limited, so one would like to keep the diffusion coefficient as close to the value of the aPM as possible. Looking at Fig. 5.23 we find that most of the local antiproton flux originates from the inner ring at 4 kpc. Although the DM halo decreases much slower than the CR source distribution in z-direction the additional antiproton sources in the halo do not increase the local antiprotons density, because the antiproton flux is always directed towards the boundary (see Eq.

2.143) [7]. This means that we are looking for a model with small diffusion coefficient in radial direction, in order to keep antiprotons from the inner ring away from the Sun, and with a vertical diffusion coefficient D_{zz} at the aPM value. Up to now we have neglected anisotropies in diffusion for practical reasons, but in fact such anisotropies are expected from the magnetic field structure of our Galaxy. The CR diffusion coefficient is basically determined by the ratio of the regular magnetic field (B) to the perturbed magnetic field (δB, the turbulences on which CRs scatter): $D \sim B^2/\delta B^2$. In regions with large magnetic field and only little turbulence CRs will follow magnetic field lines with helical trajectories. In this case the diffusion coefficient parallel to the magnetic field is large, while the diffusion coefficient perpendicular to the magnetic field is small. It has been shown in Monte Carlo simulations that this anisotropy remains even if the ratio $B^2/\delta B^2$ is small (the case of strong scattering), i.e. the diffusion coefficient along the magnetic field lines is always larger than the perpendicular diffusion coefficient (Codino & Plouin, 2007; De Marco et al., 2007). Although the absolute value of the diffusion coefficients and their ratio are unknown, because the spectrum of magnetic turbulences is not known, the *direction* of the anisotropy can be derived from the magnetic field structure of our Galaxy.

Magnetic Field Structure The magnetic field of the Milky Way is commonly modelled by two components (Han, 2004): a toroidal field following the Galaxy's spiral arms with alternating direction from arm to arm and a polodial field, which is negligible in the disk, but dominates in the halo and in the GC (see Fig. 5.25). The toroidal field dominates in the galactic disk leading to a strong transport mode in φ direction (parallel to the magnetic field in the disk along the spiral arms) while the transport in R and z direction is suppressed. The polodial field dominates in the halo leading to a preferred transport in z direction (parallel to the field in the halo) and a relatively weak transport mode in φ and R direction (see e.g. Fig. 1 in Han (2004)). Consequently, one would expect the diffusion coefficient in R direction to be smaller than the diffusion coefficient in z-direction throughout the Galaxy, while the diffusion coefficient along the spiral arms, i.e. in φ-direction dominates in the disk and becomes smaller than the diffusion coefficient in z-direction in the halo. Assuming cylindrical symmetry, diffusion in φ-direction will not change the source distribution: with no gradient in the source distribution, the gas distribution, the diffusion coefficient and all relevant fields, the problem is symmetric and no resulting flux can occur (although, of course, single particles will change places). Since

[7] Of course a single antiproton from the halo can be scattered into the plane, but at the same time more antiprotons from the plane will be scattered into the halo

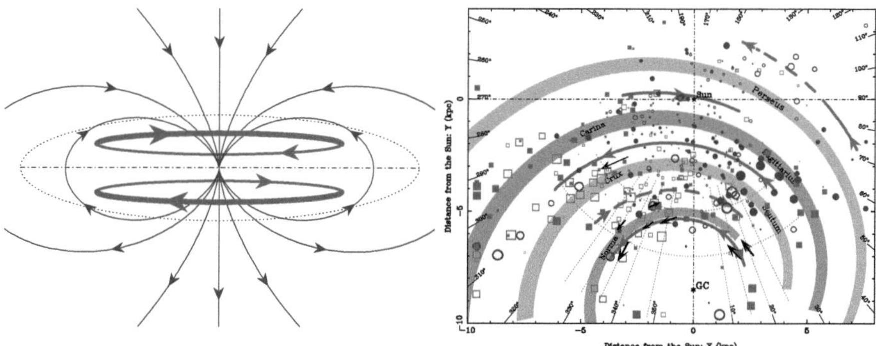

Figure 5.25: Left: Illustration of the Galactic magnetic field structure. **Right:** The galactic magnetic field in the plane as infered from pulsar rotational measures (RM). The RM distribution of pulsars projected onto the Galactic plane. Red data (squares) are newly observed, and blue (circles) are previously published. Filled symbols stand for positive RMs and open ones for negative RMs. The large-scale magnetic fields are drawn by arrows, which was inferred from RM data. Solid-line arrows stand for confirmed field structures, while dashed-line arrows stand for proposed field structures in controversy and to be confirmed. From Han (2004).

for transport along a curved path the diffusion tensor does have off-diagonal elements [8] we will consider one-dimensional transport along an infinitesimal segment in an arbitrary direction η as an example: Assuming a flat source distribution and no gradients in the diffusion coefficient as well as no convection, the transport equation 3.5 for η-direction reads:

$$\frac{\partial \Psi}{\partial t} = q(\vec{r}, t) + \frac{\partial}{\partial p} p^2 D_{pp} \frac{\partial}{\partial p} \frac{1}{p^2} \Psi - \frac{\partial}{\partial p} [\dot{p} \Psi \Psi] - \frac{1}{\tau_f} \Psi - \frac{1}{\tau_r} \Psi, \tag{5.10}$$

which means that no spatial transport along η occurs. Thus, preferred (or suppressed) transport along η will only lead to a misestimation of the amount of diffusive reacceleration by virtue or the larger (or smaller) diffusion coefficient which gives rise to a smaller (or larger) D_{pp} and to a misestimation of the momentum losses. The momentum diffusion coefficient is determined by the Alfvén velocity and the spatial diffusion coefficient as a measure for the scattering rate: $D_{pp} \sim v_a^2/D$. In the presence of anisotropic diffusion D and

[8]Due to the fact that a diffusive process always occurs along straight lines (i.e. there is no infinitesimal diffusion step along φ-direction) cylindrical coordinates do not represent a natural basis for diffusion, if an anisotropy between diffusive transport in φ and R direction is considered. A correct 3-dimensional implementation of the model therefore has to use a cartesian basis.

v_α have to be replaced by effective parameters, e.g. if $D_\varphi > D_{RR}$ the value of v_α has to be scaled with a factor $\sqrt{D_\varphi/D_{RR}}$. In addition the momentum losses through transport along φ can be considered as a reduction in v_α. Therefore, preferred transport in φ-direction can be incorporated by treating v_α as an effective parameter, which is only indirectly linked to the averaged velocity of the Alfvén waves. Remind that v_α is an effective parameter anyway, in the sense that we do not incorporate possible spatial changes in v_α which might arise from the spatial dependence of the diffusion coefficient. Finally, anisotropic diffusion in cylindrical coordinates can be modelled by the two diffusion coefficients

$$D_{RR} = \beta D^0_{RR} \left(\frac{\rho}{\rho_0}\right)^\delta ,$$
$$D_{zz} = \beta D^0_{zz} \left(\frac{\rho}{\rho_0}\right)^\delta , \tag{5.11}$$

where D^0_{RR} and D^0_{zz} are independent constants and $\delta = 0.33$, $\rho_0 = 4$ GV and an effective Alfvén velocity. From the structure of the Milky Way's magnetic field one would now expect that $D_{zz} > D_{RR}$ in the halo (where the dominant magnetic field is directed along z) and $D_{RR} \geq D_{zz}$ in the disk, where the magnetic field is directed along the spiral arms, which also have a radial component. In order to reduce the antiproton flux we require the diffusion coefficient along R to be smaller than the diffusion coefficient along z, so in the following we will assume that the anisotropy in diffusion coefficients is constant throughout the Galaxy ($D_{RR}/D_{zz} = const.$).

Results Here we estimate the impact of anisotropies in diffusion on the locally measured secondary to primary ratios and the local antiproton flux. The rigidity dependence of the diffusion coefficients is kept identical for all directions, meaning that the spectrum of turbulences is just shifted toward larger or smaller wavelengths for different directions. The anisotropy in diffusion given by $\chi_D = D_{RR}/D_{zz}$ is varied from 10^{-4} to 10^{-1}. Figure 5.26 shows the local antiproton flux for different anisotropies in diffusion for antiprotons from DMA (left) and antiprotons from CR interactions (right). A reduction in D_{RR} by a factor of 10^4, while D_{zz} is kept constant, only leads to a marginal reduction in antiprotons from DMA and to a slight increase in antiprotons from CR interactions. Since the anisotropy of 10^{-4} is already a rather extreme value, this decrease in antiprotons from DMA can be considered the maximum achievable reduction. Obviously a reduction of D_{RR} cannot

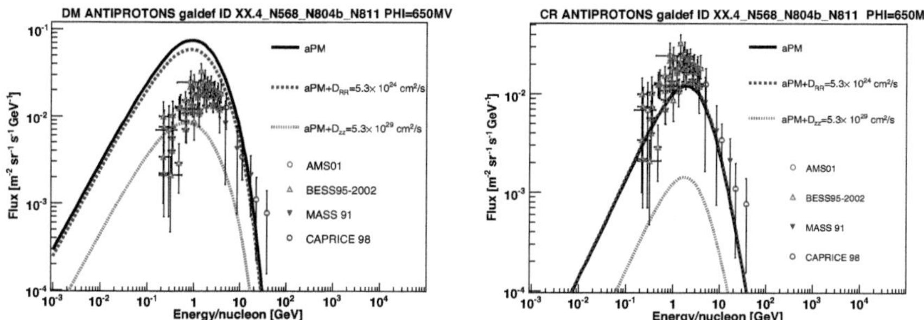

Figure 5.26: Left: Local antiproton flux from DMA for different anisotropies in diffusion. **Right:** Local antiproton flux from CR interactions for different anisotropies in diffusion.

confine the antiprotons produced in the inner ring efficiently to the source region. This is also visible from the left side of Fig. 5.27 where the radial distribution of antiprotons from DMA is shown. For the reduced diffusion coefficient in R-direction the DM antiproton distribution becomes more similar to the (squared) DM halo profile. The outer ring and the increase towards the center become more pronounced, while the reduction of local antiproton density is visible around $R = 10$ kpc, but by far not sufficient. On the other hand an increase in D_{zz} leads to a larger CR flux in z-direction. The red line in Fig. 5.26 shows that an increase in D_{zz} by only a factor of 10 is sufficient. The radial diffusion coefficient is kept at $5.3 \cdot 10^{28}\text{cm}^2/\text{s}$ in this case. Figure 5.27 impressively demonstrates that the density of antiprotons from DMA is reduced throughout the Galaxy. The relative height of the peaks due to the outer ring and the GC is not changed compared to the aPM, because D_{RR} is the same in both runs, but the overall antiproton density is reduced. At the same time the flux of antiprotons from CR interactions is greatly reduced due to the fast CR escape in z-direction. This, of course, also affects the B/C ratio, which is shown on the right side of Fig. 5.27. Clearly, in order to be compatible with the DMA interpretation of the EGRET/Fermi-LAT data, one requires a mechanism which increases, the number of secondary CRs, at least locally. This, however, constitutes a catch-22: any mechanism, which keeps the primary CRs in the Galactic disk, so that enough secondaries are produced, will also keep the antiprotons from DMA in the disk. Insofar the only option left is a local *increase* in gas density (contrary to what is required from the diffuse γ-rays). Figure 5.28 shows the minimum anisotropy in diffusion, which would lead to a

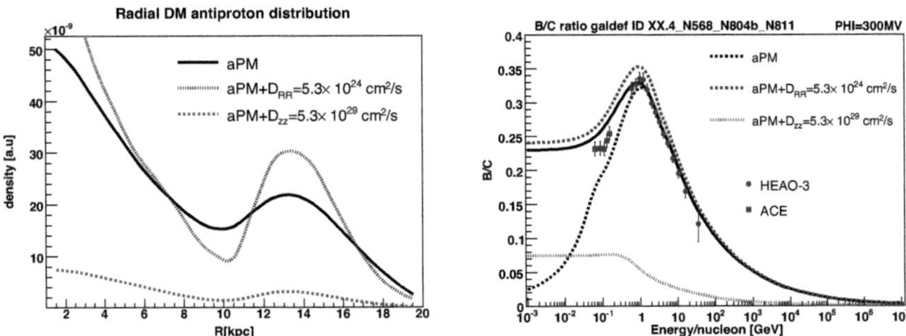

Figure 5.27: Left: Radial distribution of antiprotons from DMA for different anisotropies in diffusion. **Right:** The corresponding B/C ratio.

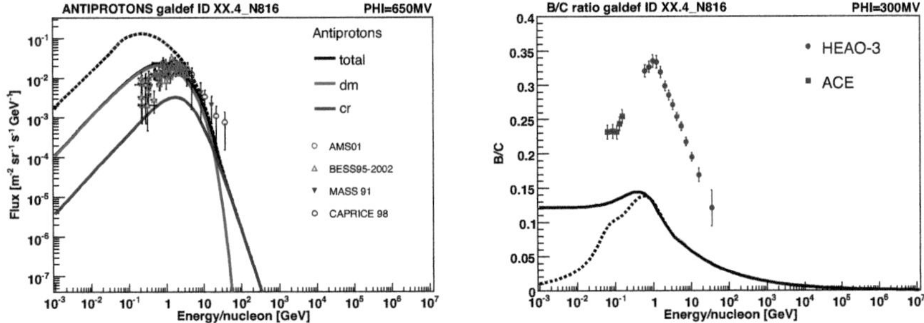

Figure 5.28: Left: Local antiproton flux in an aPM with $D_{zz} = 2.3 \cdot 10^{29}$ cm^2/s. **Right:** The corresponding B/C ratio.

local antiproton flux compatible with the data. The corresponding diffusion coefficient in z-direction is $D_{zz} = 2.3 \cdot 10^{29}$ cm^2/s, which means $\chi_D = 0.23$. Since both antiproton components are reduced by approximately the same factor, the flux from DMA still dominates, which leads to a rather soft spectrum. This could be remedied by an increase in diffusive reacceleration, however the maximum allowed Alfvén speed is constrained by B/C. The corresponding B/C ratio shown in Fig. 5.28 is still a factor 2.3 too low. To compensate this would require a significant increase in the local gas density. The Fermi data on diffuse γ-rays require a higher normalization than the local proton and electron flux and that independent of whether or not DM is invoked (see sections 5.2.3 and 5.2.4). Previously, in Section 5.3, we have seen that this is possible, if the local secondary production rate is smaller than the Galactic secondary production rate, e.g. if the Galactic local gas density is *smaller* than the global gas density. This, again, is contrary to the requirements from antiprotons from DMA.

Conclusion We have shown that with the current level of detail the local flux of antiprotons from DMA forms a tremendous constraint for the halo profile. An inner ring of DM, as expected from the EGRET analysis, will lead to a too high local antiproton density. This can be remedied by assuming an anisotropy in diffusion coefficients of the order $\chi_D = D_{RR}/D_{zz} = 0.23$. However, if this anisotropy is applied *everywhere* the local secondary production rate, as measured by the B/C ratio, cannot be reproduced. One has, however, to keep in mind that here we applied a *global* anisotropy as a simple estimate. From the spiral structure of the Galaxy as introduced in Section 5.3.4 one would expect that the diffusion coefficients vary throughout the Galaxy. In particular one would expect a larger diffusion coefficient in the spiral arms, where the CR sources are located and a higher CR density is expected. The inner DM ring roughly coincides with the two inner spiral arms, Crux and Norma, while the Sun is located in the interarm region between Sagittarius and Perseus. If the vertical diffusion coefficient D_{zz} is large in and above the spiral arms and small in the interarm regions, the antiprotons from the inner ring will escape fast into the convection zone above z_c (by virtue of a larger D_{zz}), while the local secondary production rate can be kept at a reasonable level (by virtue of a smaller D_{zz}). This way antiprotons from the inner ring will contribute less to the local antiproton flux. In order to estimate whether or not such a scenario is consistent with the constraints from γ-rays (see e.g. our discussion in 5.3.5), a detailed implementation of the spiral structure of the Milky Way is required.

5.5 Contemporary Indirect Dark Matter Searches versus Transport Model Uncertainties

In the last section we have seen that even with the additional degrees of freedom in an aPM the local antiproton flux still forms a tremendous constraint for DM models. The discussion in the previous section demonstrates the difficulties which come along with indirect DM searches in charged CRs. Diffuse γ-rays and antiprotons are just one example of many. A number of recent observations have created a blast of papers, many of them focussed on DM interpretations of the data. For most of these observations astrophysical alternatives exist, some of which even have been *expected* to be visible in CRs, like the pulsar interpretation of the PAMELA positron fraction (Blasi, 2009; Chowdhury et al., 2009; Grasso et al., 2009; Hooper et al., 2009; Profumo, 2008; Serpico, 2009; Yüksel et al., 2009). However, due to the unknown details of CR transport and the unknown parameters of astrophysical sources, DM remains a viable explanation. In this section we will review the most discussed observations and their possible interpretations and comment on them from a viewpoint of CR transport uncertainties.

We have already discussed the DMA interpretation of the INTEGRAL positron annihilation line in Section 4.3.3 and found that while CR transport can in principle explain both, the absence of an annihilation signal from positrons from SNIa from the disk and the large B/D ratio, any DMA interpretation can only account for the latter. Here we will first show how CR transport can modify the expected signal from DMA for the example of the WMAP-*haze*. We will then turn to the most discussed observations: the rising positron fraction observed by the PAMELA space-borne experiment (Adriani et al., 2009a) in combination with rather hard spectra of electrons and positrons above expected background by the FERMI satellite (Abdo et al., 2009a), the ATIC balloon experiment Chang et al. (2008) and the HESS earth-bound Cherenkov telescope (H. E. S. S. Collaboration: F. Aharonian, 2009). These excesses were not accompanied by an obvious excess in diffuse gamma rays in the halo at mid-latitude (Porter, 2009) nor in antiprotons (Adriani et al., 2009b), which has led to speculations about a new class of "leptophilic" dark matter candidates (Arkani-Hamed et al., 2009; Nomura & Thaler, 2009), which fit the data (Bergström et al., 2009).

5.5.1 The WMAP- and Fermi-*haze* as a Signature of Dark Matter

The WMAP haze (Dobler & Finkbeiner, 2008) consists of an excess of microwave emission from a small region close to the Galactic center (see the left side of Fig. 5.29), where the spectrum suddenly becomes slightly harder. WMAP covers the range from 20 GeV to 104 GeV in 5 bands with increasing badwidth (K-band (23 GHz), Ka-band (33 GHz), Q-band (41 GHz), V-band (61 GHz), W-band (94 GHz)). It was discovered in a reanalysis of the public WMAP data, where the Haslam 408 MHz map was used as a template for the synchrotron radiation (Dobler & Finkbeiner, 2008). Recently, an IC counterpart was found in the Fermi-LAT data by the same group (Dobler et al., 2009). It has been suggested that this signal could be synchrotron emission from relativistic electrons and positrons,

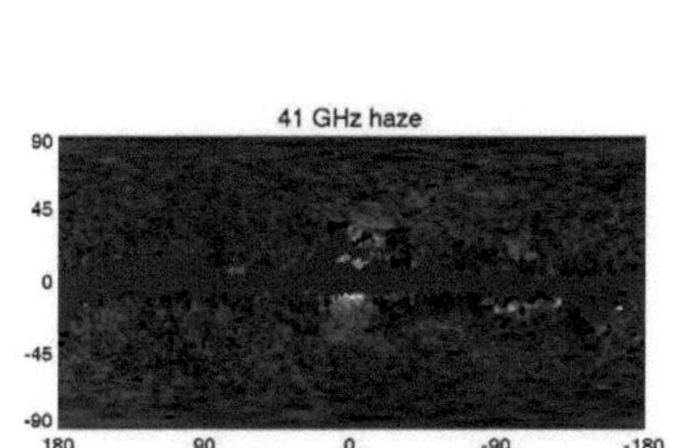

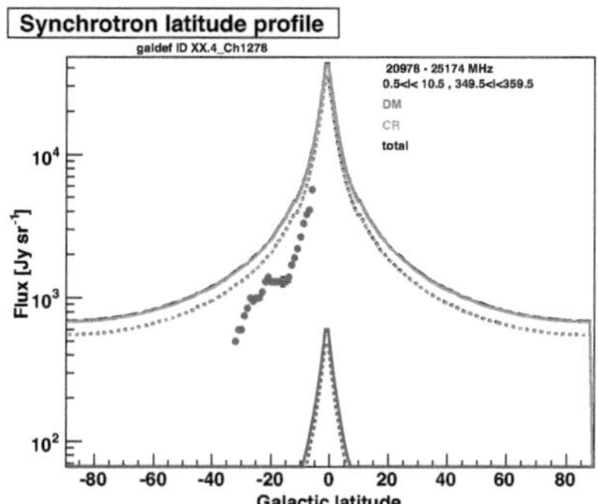

Figure 5.29: **Left:** The WMAP Q-band (41 GHz) *haze* (from Dobler et al. (2009)). **Right:** Latitude distribution of the synchrotron radiation in an aPM in the small longitude range were the haze has been measured. The top solid curve is the total flux, the red dots represent the haze (adapted from Hooper et al. (2007)) and the contribution from the DMA interpretation of the EGRET excess is shown as the red line. The dotted lines belong to the frequency bin from 17481 MHz to 20978 MHz.

possibly originating from DMA in a cuspy halo (Hooper et al., 2007). The authors used a simplified diffusion model to estimate the propagation length of the electrons and positrons from DMA. However, even in a cored profile the synchrotron radiation from the disk shows a steep increase, as we show in Fig. 5.29. Here the synchrotron emission for the 21 GHz band from CR and DM electrons in an aPM is shown. Even with a boost-factor of 27 (as expected from EGRET), the intensity of synchrotron radiation from DMA in a cored profile is too low, but the spatial shape is generally compatible with the observed *haze*. This demonstrates how difficult it is to infer the underlying halo profile from the spatial shape of the observed signal (provided that it originates from charged decay products which are subject to CR transport). Both, the existence of the WMAP-*haze* and the Fermi-*haze* and the properties of the underlying halo profile, strongly depend on the CR transport model under consideration.

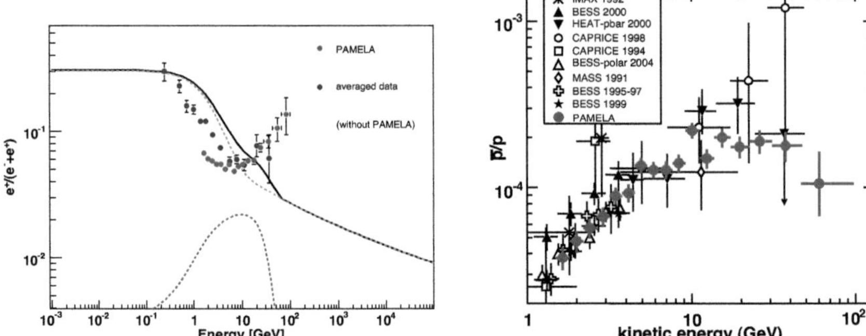

Figure 5.30: **Left:** The positron fraction measured by the PAMELA experiment compared to the prediction of the aPM. One standard deviation error bars are shown. If not visible, they lie inside the data points. The green dotted line is the contribution from CRs, the red dotted line is the contribution from DMA (for a boost factor of 10) and the black full line is the sum of both contributions. Data are from Adriani et al. (2009a). Below a few GeV the difference between the averaged data without PAMELA (blue) and the PAMELA data (red) in the low energy range is the result of a difference in solar modulation potential during the different observation periods. The solar modulation potential assumed for the model prediction has been optimized for the low-energy electron and positron data from AMS01 and therefore does not agree with the PAMELA data or the averaged data. Above a few GeV the PAMELA data roughly agree with the averaged data, but they clearly indicate a rise in the positron fraction. **Right:** The PAMELA antiproton-to-proton flux ratio from Adriani et al. (2009b) compared to contemporary measurements.

5.5.2 The "anomalous" PAMELA, ATIC, and Fermi-LAT Results on Electrons and Positrons as a Signature of Dark Matter

The positron fraction is defined as the ratio of fluxes of positrons and the sum of electrons and positrons, i.e. $e^+/(e^+ + e^-)$. The PAMELA data on the positron fraction (Adriani et al., 2009a) indicate a positron flux much harder than what is expected from CR transport (see the aPM prediction on left side of Fig. 5.30, the aPM prediction is very siilar to the predictions of isotropic models). In contrast, the antiprotons did not show any particular feature (Adriani et al., 2009b) (see the right side of Fig. 5.30). At the same time the electron spectrum as measured by ATIC (Chang et al., 2008) and PPB-BETS (Torii et al., 2008) indicates a "bump" at around 500 GeV (see e.g. Fig. 5.13 or Fig. 5.31). These observations suggest an additional hard positron and electron component which dominates the local spectra for higher energies. Both results have been interpreted as a possible signal from DMA by many authors (for an extensive compilation of references see Profumo (2008)). The bump in the electron spectrum has led to speculations about new physics, especially the possibility of the annihilation of Kaluza-Klein type WIMPs. Since the excess stops around 800 GeV, this would require WIMP masses in this range. It is interesting to note that ATIC measures an increase above the index from spectrometer experiments for both the electron and proton/nuclei spectra (Panov et al., 2006). Just at the edge of acceptance the spectra then fall sharply, which may point to background from heavy nuclei increasing towards higher energies. In fact, the ATIC-bump was not confirmed by data from the Fermi telescope (Abdo et al., 2009a). The Fermi data show a smooth spectrum as can be seen e.g. from Fig. 5.31. The spectrum up to 1 TeV is well described by a power law proportional to $E^{-3.0}$ in agreement with data from the HESS experiment as shown in Fig. 5.31. The HESS data also show no indication of a structure in the electron spectrum, but rather a power-law spectrum with a spectral index of $3.0 \pm 0.1(stat.) \pm 0.3(syst.)$, which steepens at about 1 TeV.

CR positrons are produced by the decay of positively charged pions produced by inelastic collisions of CRs with the gas in the disk. Electrons mainly originate from SNRs and the fraction of electrons produced by the decays of negative pions from CR interactions is small compared to the primary electrons from SNRs. In the absence of additional sources, the local positron spectrum is therefore related to the local spectrum of the nuclei and the energy losses, while the local electron spectrum is determined by the electron injection index and the electron energy losses. Positrons from DMA are produced mainly by the decays of positively charged pions produced after the hadronization of the quarks. The contribution from DMA in the local $e^+/(e^+ + e^-)$-fraction is highly uncertain and does depend on the transport model. In an aPM the contribution of DMA from a halo profile as preferred by EGRET and a boost-factor of 10 is small, as shown in Fig. 5.30.

Since the observed rise in the PAMELA positron fraction is not accompanied by a rise in antiprotons (see the right side of Fig. 5.30), any DMA interpretation of the data requires a leptophilic WIMP. These WIMPS decay into new states, which are too light to produce antiprotons, i.e. typical WIMP masses are below 1 GeV. For example, Nomura & Thaler (2009) proposed DM fermions decaying into an axion and a scalar with the latter decaying again to axions. For an axion mass in the range 360 - 800 MeV antiproton production is forbidden.

Several other DM candidates have been discussed in the literature (see e.g. the list of references in Profumo (2008) or the review by de Boer (2009)). However, even if a certain candidate matches the constraints from local antiprotons and diffuse γ-rays (which would require a rather fine-tuned DM model), the energetic e^+e^- pairs should be visible in synchrotron radiation, especially in the GC. All this strongly points to a local point source or a group of local point sources, such as pulsars, as has been suggested by many authors (Aharonian et al., 1995; Coutu et al., 1999; Hooper et al., 2009; Profumo, 2008; Serpico, 2009). In the following we will discuss the most promising astrophysical ideas to explain the PAMELA data.

Astrophysical Explanations for PAMELA

The increase in the positron fraction cannot be explained by current propagation models in a consistent picture as demonstrated in Fig. 5.30. Blasi (2009) suggested that additional positrons could be generated and accelerated in old SNRs which are located inside dense molecular clouds. Distant SNRs generate a softer spectrum than local SNRs due to positron energy losses during propagation. Provided that there are enough SNRs at a distance of 1-2 kpc, this mechanism could explain the cut-off at about 1 TeV in agreement with the HESS data (H. E. S. S. Collaboration: F. Aharonian, 2009). If the rise in the positron fraction is due to the acceleration of secondary positrons in SNRs the same mechanism should work for antiprotons. Consequently one would expect a corresponding rise in the antiproton/proton ratio, which is currently not observed (see Fig. 5.30). However, this increase may happen above 100 GeV, as shown by Blasi & Serpico (2009). Other astrophysical explanations are positrons from pulsars and electrons from local sources, which will be discussed in the following.

Pulsars Pulsars are rapidly spinning, magnetized neutron stars. They emit electromagnetic radiation along the magnetic poles, which can be observed when the emission cone strikes the Earth. In the strong magnetic fields in the polar region of a pulsar photons are created by synchrotron emission. These photons can generate electrons and positions by interaction with the magnetic field, which in turn produce synchrotron radiation. This way a large number of high-energy positrons and electrons, but no antiprotons, can be produced. The energy and the amount of escaping positrons and electrons depend on the specific properties of the pulsar, which are unknown: especially young pulsars are surrounded by nebulae (the remnant of the SN explosion creating the pulsar). These nebulae

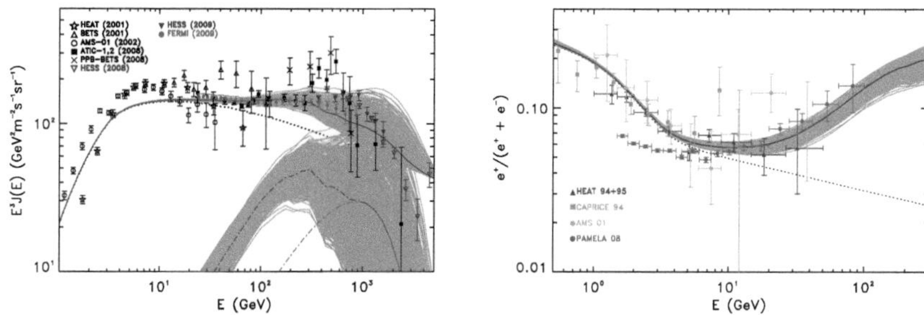

Figure 5.31: Electron plus positron spectrum (**left**) and positron fraction (**right**) from multiple pulsars plus the Galactic component with experimental data (dotted line). Each gray line represents the sum of all pulsars for a particular combination of pulsar parameters. The blue dot-dashed (pulsars only) and blue solid lines (pulsars + Galactic CR component) correspond to a representative choice among the set of possible realizations. From Grasso et al. (2009).

are thought to be the prime acceleration region for CRs, because a first order Fermi acceleration process is expected to occur in the expanding SN shells. In addition to the directed shock fronts there are undirected magnetic turbulences in the nebulae, which can trap the electrons and positrons in their magnetic fields for about 10^5 years. Therefore one has to consider pulsars older than this. Assuming both, the escape rate and the number of old and nearby pulsars to be free parameters, pulsars can nicely explain the PAMELA and Fermi data, as shown in Fig. 5.31 (Grasso et al., 2009). If only a few nearby sources contribute, an anisotropy in the positron flux above 10 GeV is expected: Although positrons and electrons that leave the pulsar nebulae will loose any directional information quickly due to resonant scattering, a larger electron and positron flux is expected from the pulsar region (see Eq. 2.143). Since the local pulsar candidates are located at larger Galactocentric radii than the Sun, the anisotropy in the electron and positron fluxes, which is generally compatible with a CR flux out of the source region toward the outer Galaxy, will decrease or might even change sign above a few hundred GeV. A review about pulsars and how they can explain the data was given by Profumo (2008).

Local Sources The spiral structure of our Galaxy, introduced in Section 5.3.4, might also help to explain the Fermi and PAMELA data on electrons and positrons. Sagittarius and

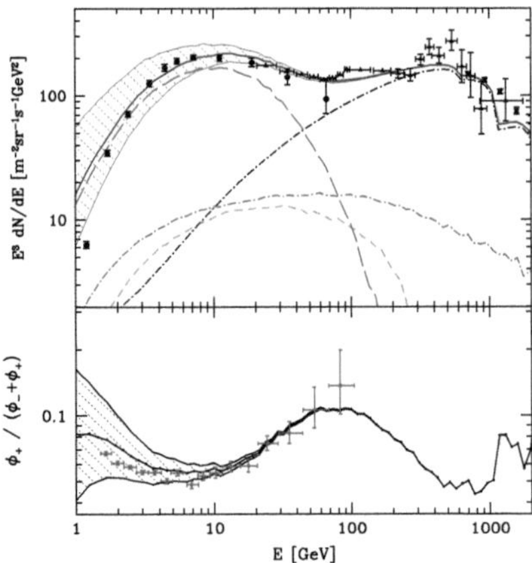

Figure 5.32: **Top**: Electron and positron spectra for primary arm electrons (long dashed purple), primary disk electrons with nearby sources excluded (short dashed green), nearby SNRs (dot-dashed black), secondary positrons (dot-dashed red), and their sum (blue). The hatched region describes the solar modulation range (from 200 MV to 1200 MV). **Bottom**: The positron fraction for the model shown in the top panel. From Shaviv et al. (2009)

Perseus, the nearest arms, are located at distances of around 1 kpc and 2-3 kpc, respectively (see Fig. 5.22). Since CRs produced in these arms have to travel significant distances, they will suffer significant energy losses and consequently a large fraction of electrons with a relatively soft spectrum will arive at Earth. Electrons produced in the Galactic disk will show a somewhat harder spectrum and very local sources, which are possibly located just outside our LB, will contribute predominantly at the highest energies. Assuming an arbitrary normalization of these three components, they can describe the Fermi electron spectrum resonably well, as shown in Fig. 5.32 (Shaviv et al., 2009).

Positrons are produced from CR interactions in the gas. Although the gas is far from being homogeneous (the gas also follows the spiral arms and there can be "local positron sources" in the form of gas clumps) the authors of Shaviv et al. (2009) assumed that positrons come from all distances and have a smooth distribution. In their model the resulting positron fraction first increases and then decreases, similarly to what would be expected from dark matter annihilation, but different from most other astrophysical explanations.

5.6 A Comment on Simplicity and Complexity of Models

By the time this is written there are more than a hundred papers on DM interpretations of the PAMELA, Fermi, ATIC, INTEGRAL and WMAP results available on the ArXiv and only a small fraction includes estimates of a possible contribution from astrophysical sources. It may appear as a shift of paradigm, that the *unknown* sources (invoking new physics) receive much more attention than the *known* sources. Unfortunately both, astrophysical explanations and explanations invoking new physics, depend on a number of free parameters which can be fitted to explain all current observations reasonably well.
In his paper entitled "Dissecting Pamela (and ATIC) with Occam's Razor: existing, well-known Pulsars naturally account for the 'anomalous' Cosmic-Ray Electron and Positron Data" Profumo (2008) reminded the reader, that in the presence of two competing theories with comparable prediction, the simpler theory is the better choice. Complexity and simplicity are of course perspectives of notion and one must not forget that the existence of DM is a well-established fact, confirmed by a variety of independend observations, such as the rotation curves of galaxies and clusters of galaxies, the primordial density fluctuations, observed as fluctuations in the CMB radiation by COBE, WMAP and Planck (which is currenly taking data), the gravitational lensing of galaxy clusters and many others. If DM is a thermal relic from the big bang, it has to annihilate into standard model particles which can be detected. The exact annihilation rate depends on the substructure of the DM halo, which unfortunately is only poorly constrained by models for Galactic structure formation. Therefore the absolute strength of a possible DM signal constitutes a free parameter for any model and any DMA signal may as well dissapper in the CR background. Nevertheless, a complete neglection of a possible contribution from DMA would constitute an incorrect hypothesis and if DM is indeed visible in CRs this will inevitably lead to false conclusions. It is the opinion of the author that the large number of DM interpretations concerning the PAMELA, Fermi, ATIC, INTEGRAL and WMAP data are to be seen in this context.
It is one of the greatest challenges of physics to simultaneously envision multiple possibilities of how nature could be and to keep track of what we actually know and how we know it. Until the definite discovery of DM, which has to be accompanied by the extraordinary evidence required by an extraordinary claim, particle physicists, astrophysicists and cosmologists can learn a lot from each other.

Chapter 6

Summary and Outlook

Our information on cosmic ray transport largely stems from two entirely different sources: charged cosmic rays provide information about the local cosmic ray densities and γ-rays provide information about the interstellar environment. Cosmic ray transport in our Galaxy has been successfully described by isotropic diffusion models, where the locally observed cosmic ray fluxes can be used in conjunction with the interstellar data on diffuse gamma rays to constrain the transport parameters. It has been shown by several authors that it is possible to describe the local fluxes of charged cosmic rays, including cosmic ray clocks such as the $^{10}Be/^{9}Be$ ratio and the secondary fluxes as well as the Galactic γ-rays up to 1 GeV in such a model. Since the knowledge of the exact setup of our Galaxy is limited, the predictive power of any model for cosmic ray transport is narrowed and no set of transport parameters can be considered unique. An example for this degeneracy is the size of the diffusion region which crucially determines the diffusion coefficient.

One of the most advanced programs providing a numerical solution to the diffusion equation is the publicly available GALPROP code. The basic parameters are the injection spectrum, diffusion coefficient, convection velocity, Alfvén velocity and the size of the halo. Together with the diffusion coefficient the latter determines the cosmic ray residence time in the Galaxy, since as soon as they pass the border, they are assumed to escape to outer space. By tuning these parameters to the secondary/primary ratio and the unstable/stable ratio one obtains a self-consistent propagation model of our Galaxy. The amount of secondary cosmic ray particles and Galactic γ-rays are described by the cross sections of the interactions of the primary and secondary cosmic ray with the gas of the disk using a network with more than 2000 cross sections. Since γ-ray emission and cosmic ray densities are determined self-consistently for all regions of the Milky Way, the GALPROP models are widely-used in the context of astrophysical research and in the context of indirect Galactic dark matter searches. The Fermi-LAT diffuse emission models, for example, are currently based on the predictions of the isotropic GALPROP models and virtually all studies of a possible dark matter contribution in cosmic rays use GALPROP in order to estimate the contribution from secondary positrons. However, the GALPROP models can only allow for isotropic and homogeneous cosmic ray transport. In particular, the convection velocity is limited to a few tens of km/s in the halo. For the equidistant grid used in GALPROP the spatial resolution is strictly limited by the available memory resources. This smears out small scales structures, such as the Local Bubble, a low density region surrounding our Sun, or the spiral structure of the Milky Way, and makes them unavailable to studies.

In the past years new observations have increased our knowledge about the transport processes in our Galaxy and, in particular, in the Sun's local environment. At the same time more accurate data on the local cosmic ray fluxes have demonstrated the necessity for a more detailed transport model. Here we argued that one has to resort to a more detailed model in order to be compatible with all observations: With the insight that the cosmic ray pressure can launch Galactic winds in the Milky Way (as deduced from the X-ray data from the ROSAT satellite) the assumption of isotropic cosmic ray transport has to be dropped. Simultaneously, the observation of a rising positron fraction can be explained only if local sources, such as pulsars, are taken into account or if an additional contribution from dark matter annihilation is assumed.
In this thesis a new model for Galactic cosmic ray transport, which allows for significant convective transport compatible with the ROSAT observations, is presented. The model was realized by modifying the publicly available GALPROP code, which up to now allowed only isotropic transport and spatially constant transport parameters. The GALPROP code was modified in the following way:

- the Galactic winds were assumed to be proportional to the cosmic ray source distribution, which was taken to be the supernova remnant distribution
- the mean free path of cosmic rays - and therefore the diffusion coefficient - in the halo was assumed to increase linear with the distance from the disk.

Fixing the magnitude of the convection speed to the wind speeds suggested by the ROSAT data, the increase in diffusion coefficient in the halo can be fitted from the amount of secondary production (from the B/C ratio) and the residence time of cosmic rays (from the cosmic clocks, in this case the $^{10}Be/^{9}Be$ ratio). It is shown that such a model is consistent with all available cosmic ray data, including not only the ROSAT data on convective winds, but also the large bulge-over-disk ratio of the positron annihilation line as observed by the INTEGRAL satellite. In an anisotropic model disk positrons are efficiently transported to the halo, were they find no electrons to annihilate with.

A novel and important feature of the anisotropic propagation model is the smooth transition to free escape of cosmic rays, because of the increase in mean free path with increasing distances from the disk. Therefore the boundary condition can be moved to infinity in contrast to isotropic propagation models, where the boundary condition is fine-tuned to get the correct residence time of cosmic rays inside the Galaxy. This features allows for the first time a realistic modelling of cosmic ray escape from the Galaxy. Given the fact that all transport parameters are directly or indirectly constrained by the cosmic ray flux towards the boundary, the independence of the cosmic ray flux from the boundary condition is a great improvement.

With increasing level of data accuracy modelling of the substructures in the ISM and variations in the transport parameters becomes more and more important. An example for this are the preliminary Fermi data on diffuse γ-rays which require a somewhat different normalization than the local protons and electrons. The modified GALPROP version presented here is the first tool capable of estimating the impact of Galactic structures, such as the Local Bubble or the spiral structure of our Galaxy, which might help to explain this discrepancy. We have shown that in an anisotropic transport model the Fermi data are in principle compatible with the local cosmic ray density, if structures like the Local Bubble and the spiral arms of the Galaxy are taken into account. At the same time a more detailed model for cosmic ray transport also allows us to improve the constraints on a possible contribution from dark matter annihilation in charged cosmic rays or diffuse γ-rays. We found the dark matter interpretation of the EGRET and Fermi data to be tightly constrained by the local antiproton flux, while the PAMELA and Fermi data on electrons and positrons allow for a variety of dark matter and astrophysical explanations. We have demonstrated at the example of the WMAP-*haze*, the INTEGRAL positron signal and the rise in the positron fraction as observed by PAMELA, that any attempt to disentangle a potential dark matter annihilation signal from astrophysics requires deep understanding of the conventional astrophysics background.

Currently the PAMELA and Fermi data on electrons and positrons allow for a variety of

different explanations: positrons and electrons from local supernova remnants and electrons from the closest spiral arm can describe the rise in the positron fraction and the electron spectrum just as well as an additional contribution from dark matter annihilation or positrons and electrons from local pulsars. In 2010 shuttle flight STS-134 will launch and bring the AMS-02 detector to the International Space Station. For the first time cosmic ray data between 1 GeV and 1 TeV will be measured simultaneously by the same detector. AMS-02 will cover an energy range which is currently unavailable. In particular, AMS-02 will extend the positron measurements above 100 GeV, thus allowing us to test the pulsar hypothesis, which predicts a rising positron fraction beyond this energy. The dark matter annihilation and supernova remnant hypothesis on the other hand, predict a decrease in the positron fraction at about 100 GeV. At the same time an accurate measurement of the antiproton spectrum at high energies might help to support or rule out the supernova remnant hypothesis, which predicts a rise in the local antiproton flux above 100 GeV. The behavior of the electron, positron and especially antiproton spectrum in the AMS-02 data will hopefully help to falsify some of the hypothesis that are currently under discussion.
The anisotropic propagation model developed in the framework of this thesis is the most consistent transport model currently at our hands and will be most helpful in these future studies. The results shown in this work have been derived in a 2-dimensional model with cylindrical symmetry. A full 3D implementation of this model will allow us to better estimate the uncertainties in the local electron, positron and antiproton fluxes originating from structures like the Local Bubble and the spiral arms, as well as a possible contribution from dark matter annihilation.

Appendix A

Energy Losses

A.1 Bremsstrahlung

For electrons and positrons the energy losses due to bremsstrahlung in the ISM become relevant. **Electron-proton** bremmstrahlung in a cold plasma is goverened by the equation (von Stickforth, 1961)

$$\begin{aligned}\left(\frac{dE}{dt}\right)_{ep} &= -\frac{2}{3}\alpha_f r_e^2 m_e c^2 Z^2 n \cdot \\ &\cdot \begin{cases} 8\gamma\beta[1-0.25(\gamma-1)+0.44935(\gamma-1)^2-0.16577(\gamma-1)^3], & \gamma \leq 2; \\ \beta^{-1}[6\gamma\ln(2\gamma)-2\gamma-0.29], & \gamma \geq 2. \end{cases}\end{aligned} \tag{A.1}$$

For the **electron-electron** bremsstrahlung one can obtain (Haug, 1975; Moskalenko & Jourdain, 1997)

$$\left(\frac{dE}{dt}\right)_{ee} = -\frac{1}{2}\alpha_f r_e^2 m_e c^2 Z n \beta\gamma^* Q_{cm}(\gamma^*), \tag{A.2}$$

where

$$Q_{cm}(\gamma^*) = 8\frac{p^{*2}}{\gamma^*}\left[1-\frac{4p^8}{3\gamma^*}+\frac{2}{3}\left(2+\frac{p^{*2}}{\gamma^{*2}}\right)\ln(p^*+\gamma^*)\right],$$
$$\gamma^* = \sqrt{(\gamma+1)/2},\ p^* = \sqrt{(\gamma-1)/2},$$

and the asteriks denotes center-of-mass variables. The **total** bremsstrahlung losses in the ionized gas are given by the sum $(dE/dt)_{BI} = (dE/dt)_{ep} + (dE/dt)_{ee}$. A good approximantion is given by (Ginzburg, 1979, p.408)

$$\left(\frac{dE}{dt}\right)_{BI} = -4\alpha_f r_e^2 m_e c^2 Z(Z+1) n E \left[\ln(2\gamma) - \frac{1}{3}\right] \tag{A.3}$$

Bremsstrahlung energy losses in neutral gas can be optained by integration over the bremsstrahlung luminosity (Koch & Motz, 1959)

$$\left(\frac{dE}{dt}\right)_{B0} = -c\beta \sum_{s=H,He} n_s \int dk\, k \frac{d\sigma_s}{dk}. \tag{A.4}$$

A suitable approximation for equation A.4 is given by (Ginzburg, 1979, p.386,409)

$$\left(\frac{dE}{dt}\right)_{B0} = \begin{cases} -4\alpha_f r_e^2 m_e c^2 E \left[\ln(2\gamma) - \frac{1}{3}\right] \sum\limits_{s=H,He} n_s Z_s (Z_s+1), & \gamma \leq 100; \\ -cE \sum\limits_{s=H,He} \frac{n_s M_s}{T_s}, & \gamma \geq 800, \end{cases} \tag{A.5}$$

with a linear connection in between. Here M_s is the atomic mass and T_s is the radiation length ($T_H \simeq 62.8$ g/cm^2, $T_{He} \simeq 93.1$ g/cm^2).

A.2 Compton losses

The Compton energy losses are calculated using the Klein-Nishina cross section (Jones, 1965; Moskalenko & Jourdain, 1997)

$$\left(\frac{dE}{dt}\right)_{Cpt} = \frac{\pi r_e^2 m_e c^2 c}{2\gamma^2 \beta} \int_0^\infty d\omega\, f_\gamma(\omega)[S(\gamma,\omega,k^+) - S(\gamma,\omega,k^-)], \tag{A.6}$$

where the background photon distribution, $f_\gamma(\omega)$, is normalized to the photon number density as $n_\gamma = \int d\omega\, \omega^2 f_\gamma(\omega)$, ω is the energy of the background photon in the electron rest-mass frame, $k^\pm$ is given by $k^\pm = \omega\gamma(1 \pm \beta)$ and

$$\begin{aligned} S(\gamma,\omega,k) = \omega \Bigg\{ & \left(k + \frac{31}{6} + \frac{5}{k} + \frac{3}{2k^2}\right) \ln(2k+1) - \frac{11}{6}k - \frac{3}{k} + \frac{1}{12(2k+1)} + \\ & + \frac{1}{12(2k+1)^2} + Li_2(-2k) \Bigg\} - \gamma \left\{ \left(k + 6 + \frac{3}{k}\right) \ln(2k+1) - \right. \\ & \left. - \frac{11}{6}k + \frac{1}{4(2k+1) - \frac{1}{12(2k+1)^2}} + 2Li_2(-2k) \right\}, \end{aligned} \tag{A.7}$$

and Li_2 is the dilogarithm:

$$Li_2 = -\int_0^{-2k} dx \, \frac{1}{x} \ln(1-x) \tag{A.8}$$

$$= \begin{cases} \sum_{i=1}^{\infty} (-2k)^i/i^2, k \leq 0.2; \\ -1.6449341 + \frac{1}{2}\ln^2(2k+1) - \ln(2k+1)\ln(2k) + \sum_{i=1}^{\infty} i^{-1}(2k+1)^{-1}, k \geq 0.2. \end{cases}$$

A.3 Synchrotron losses

Synchrotron energy losses are given by

$$\left(\frac{dE}{dt}\right)_S = -\frac{32}{9}\pi r_e^2 c U_B \gamma^2 \beta^2, \tag{A.9}$$

where $U_B = \frac{H^2}{8\pi}$ is the energy density of the *random* magnetic field.

A.4 Ionization Losses

Ionization losses in the ISM can be written as:

$$\left(\frac{dE}{dt}\right)_I (\beta \leq \beta_0) = -2\pi r_e^2 m_e c^2 Z^2 \frac{1}{\beta} \sum_{s=H,He} n_s [B_s + B'(\alpha_f z/\beta)] \tag{A.10}$$

(Mannheim & Schlickeiser, 1994, Eq. 4.24), where α_f is the fine structure constant, n_s is the number density of the corresponding species in the ISM, $\beta_0 = 1.4e^2/\hbar c - 0.01$ is the characteristic velocity of the electrons in hydrogen, and

$$B_s = \left[\ln\left(\frac{2m_e c^2 \beta^2 \gamma^2 Q_{max}}{\tilde{I}_s^2}\right) - 2\beta^2 - \frac{2C_s}{z_s} - \delta_s\right], \tag{A.11}$$

where γ is the Lorentz factor of the ion. The largest possible energy transfer from the incident particle to the atmic electron is defined by kinematics

$$Q_{max} \approx \frac{2m_e c^2 \beta^2 \gamma^2}{1 + [2\gamma m_e/M]}, \tag{A.12}$$

where $M >> m_e$ is the nucleon mass and $\tilde{I}_s$denotes the geometric mean of all ionization and exitation potentials of the atom. The values $\tilde{I}_H = 19$ eVand $\tilde{I}_He = 44$ eV are given in Mannheim & Schlickeiser (1994). The shell correction term C_s/z_s, the density correction term δ_s and the B' correction term (for large Z or small β) in equation A.10 can be neglected for this purpose, so that we end up with

$$\left(\frac{dE}{dt}\right)_I (\beta \leq \beta_0) = -2\pi r_e^2 m_e c^2 Z^2 \frac{1}{\beta} \sum_{s=H,He} n_s \left[\ln\left(\frac{2m_e c^2 \beta^2 \gamma^2 Q_{max}}{\tilde{I}_s^2}\right) - 2\beta^2\right], \quad \text{(A.13)}$$

for the nucleon ionization losses.

For electron ionization losses in a medium of neutral hydrogen and helium the Bethe-Bloch formula has to be applied

$$\left(\frac{dE}{dt}\right)_I = -2\pi r_e^2 m_e c^2 \frac{1}{\beta} \sum_{s=H,He} Z_s n_s \left[\ln\left\{\frac{(\gamma-1)\beta^2 E^2}{2I_s^2}\right\} + \frac{1}{8}\right], \quad \text{(A.14)}$$

where Z_s is the nucleus charge, n_s is the gas number density, I_s is the ionization potential ($I_H = 13.6$eV and $I_{He} = 24.6$eV) and E is the total electron energy.

A.5 Coulomb Scattering

Coulomb collisions of **nuclei** in a completely ionized plasma are dominated by scattering off the thermal electrons. The corrsponding energy losses are given by Mannheim & Schlickeiser (1994, Eqs. 4.16, 4.11),

$$\left(\frac{dE}{dt}\right)_{Coul} \approx -4\pi r_e^2 m_e c^2 Z^2 n_e \ln\Lambda \frac{\beta^2}{x_m^3 + \beta^3}, \quad \text{(A.15)}$$

where r_e is the classical electron radius, m_e is the electron rest mass, Z is the projectile nucleon charge, n_e is the electron number density in the plasma, $x_m \equiv (3\sqrt{\pi}/4)^{1/3} \times \sqrt{2kT_e/m_e c^2}$, and T_e is the electron temperature. The Coulomb logarithm $\ln\Lambda$ in the cold plasma limit is given by (e.g. Dermer (1985))

$$\ln\Lambda \approx \frac{1}{2}\left(\frac{m_e^2 c^4}{\pi r_e \hbar^2 c^2 n_e} \cdot \frac{M\gamma^2\beta^4}{M + 2\gamma m_e}\right), \quad \text{(A.16)}$$

where M is the nucleon mass. For appropiate number densities, $n_e \sim 10^{-1} - 10^{-3}$ cm^{-3}, and total energies, $E \sim 10^3 - 10^4$MeV, the typical value of the Coulomb logarithm $\ln\Lambda$

lies in the range of $\sim 40-50$, instead of the value of 20, which is usually adopted.
For **electrons**, the Coulomb energy losses in the fully ionized medium in the cold plasma limit are described by Ginzburg (1979)

$$\left(\frac{dE}{dt}\right)_{Coul} = -2\pi r_e^2 m_e c^2 n_e \frac{1}{\beta}\left[\ln\left(\frac{E m_e c^2}{4\pi r_e \hbar^2 c^2 n_e}\right) - \frac{3}{4}\right] \tag{A.17}$$

where n_e ist the electron number density. For an accurate treatment of the electron energy losses in a plasma of arbitrary temperature see e.g. Dermer (1985); Moskalenko & Jourdain (1997).

A.6 Ineleastic Scattering

CRs crossing the thin Galactic disk it may undergo nuclear interactions with the interstellar hydrogen or helium. These encounters can result in inelastic scattering with the result that the paren nuceus is destroyed and new CR secondaries are created. The fragmentation of the initial CR nucleus is goverend by the total cross section, while the production of the doughter nuclei is given by the branching ratio for each channel. In the first case we talk about *fragmentation*, the second case is called *spallation*. In a diffusion equation fragmentation can be treated by associating a fragmentation rate $1/\tau_f$ to the cross sections. For CR transport the most important aspect of spallation is that nuclei, which are not produced in SN explosions can be created. The interstellar medium is mainly composed of H and He, so that the most relevant contribution to secondaries comes from reactions such as

$$n_1 + (p, He) \to n_2 + X, \tag{A.18}$$

where n_1 stands for some kind of primary nucleus while n_2 is the secondary produced in the interaction. Also heavier components of the ISM can contribute to secondary production, so in fact we have

$$(n_1 + (p, He, CNO, Fe) \to n_2 + X) \tag{A.19}$$

From A.18 and A.19 it follows that the source term for secondary nucleons can be written as

$$q_j(\vec{r},p) = \beta c \sum_i f_i^{prim}(\vec{r},p) \cdot \left[\sigma_{ij}^{(H)}(p) n_H(\vec{r}) + \sigma_{ij}^{(He)}(p) n_{He}(\vec{r}) + \sigma_{ij}^{(CNO)}(p) n_{CNO}(\vec{r}) + \right. \tag{A.20}$$
$$\left. + \sigma_{ij}^{(Fe)}(p) n_{Fe}(\vec{r})\right],$$

where f_i^{prim} is the primaries density of the parent nucleus i, $\sigma_{ij}^{(H,He,CNO,Fe)}$ is the cross section for the production of nucleus j from nucleus i that scatters on H, He, CNO, FE. One can disregard the i index by introducing the weighted production cross sections $\sigma_j^{(X)} = \sum_i f_i^{prim} \sigma_{ij}^{(X)}$ for the daughter j-nucleus produced on the X-nucleus target, so that we are left with

$$q_j(\vec{r}, p) = \beta c \cdot \left[\sigma_j^{(H)}(p) n_H(\vec{r}) + \sigma_j^{(He)}(p) n_{He}(\vec{r}) + \sigma_j^{(CNO)}(p) n_{CNO}(\vec{r}) + \right. \\ \left. + \sigma_j^{(Fe)}(p) n_{Fe}(\vec{r})\right], \tag{A.21}$$

A comparision of the mean time of each reaction on a target (X) $\tau = \beta c \sigma_{ij}^{(X)} n_{(X)}$and the typical propagation time of CRs immediately gives an estimate of the relevance of the different scattering processes. In nuclear reactions in the interstellar medium, the kinetic energy per nucleon is approximately conserved so that all the information is contained in cross sections. The isotopic cross section database in GALPROP currently consists of more than 2000 points collected from sources published in 1969-1999.

A.7 Radioactive Decay

Radioactive instable isotopes in CRs are created both, by fragmentation of heavier nuclei (e.g. ^{10}Be) and directly in the CR sources (e.g. ^{26}Al). For a large population of nuclei we can use the statisical approximation

$$N(t) = N_0 e^{t/\ln 2 \tau_{1/2}} \tag{A.22}$$

in order to describe radioactive decays. Here N_0 is the intial population and $\tau_{1/2}$ is the half lifetime. If we consider a frame moving with relativistic speeds, the half-lifetime becomes $\tau_{1/2} \rightarrow \gamma \tau_{1/2}$.

So far the problem is well defined and easy to solve once the half-lifetimes associated to each nucleus are given. Unfortuanetly, different decay modes are affected in different ways by CR transport, while the half-lifetime is often a a combination of all decay modes. In the context of CR propagation, the most relevant decay processes are β-decay and K-capture, where the nucleus captures a K-shell electron. Naturally β-decay does not depend on a specific environment, while for K-capture an attached electron is necessary. Since the ISM is very poor in electrons the lifetime of a K-capture isotope can be significantly longer than the ones measured on Earth. The importance of unstable elements resides in the possibility of measuring the average local age of CRs. Pioneering work on this subject focussed on ^{10}Be (Garcia-Munoz et al., 1977; Hagen et al., 1977; Webber et al., 1977) followed by heavier elemts like ^{26}Al (Freier et al., 1980), ^{36}Cl (Wiedenbeck, 1985) and ^{54}Mn (Leske, 1993). It truns out that the typical age of a CR is about 10 – 20Myr.This value can be used to select the interesting unstable nuclei for cosmic ray physics once the γ-factor that enhances the lifetime of the nuclei is taken into account. In Donato et al. (2002) a complete list of such a nuclei is given. Only three purely β-decay unstable elements have a half-life within the interval 1 kyr-100 Myr as reported in Table A.1 The transition from Fe to Ni, includes the transition from Fe to Co as an intermediate step with a half-life of 1.5 Myr while transition from Co to Ni is immediate from a cosmic ray point of view ($\tau_{1/2} \sim 5$ yr). Such short-lived intermediate steps do not have to be propagated and the complete decay chain can be executed at once.

Z	Nucleus	Daughter	$t_{1/2}^{\text{unit}}$(error)
4	$^{10}_{4}$Be	$^{10}_{5}$B	$1.51^{\text{Myr}}(0.06)$
6	$^{14}_{6}$C	$^{14}_{7}$N	$5.73^{\text{kyr}}(0.04)$
26	$^{60}_{26}$Fe	$(^{60}_{27}\text{Co} \xrightarrow{\beta-})^{60}_{28}$Ni	$1.5^{\text{Myr}}(0.3)$

Table A.1: Pure β unstable isotopes (1 kyr $< t_{1/2} <$ 100 Myr) from Donato et al. (2002)

The interesting elements that undergo purely electronic capture are summed in Table A.2. Another set of nuclei that are worth to be considered, show a mixed electronic capture and β decay that means that the half-lives associated to the two channels are comparable. A list of these nuclei is presented in Table A.3. An obvious application of unstable nuclei in the context of cosmic ray physics, is the evaluation of the halo height. Strong & Moskalenko (1998) used the ratio ^{10}Be/^{9}Be to gain information about the halo in connection with the Ulysses data (Connell et al., 1998). Further improvement was presented in Strong & Moskalenko (2001) where the ACE data were added to obtain the more robust estimate $z_h = 3{-}7$kpc. Donato et al. (2002) pointed out that the most probable distance $L = \sqrt{D\gamma\tau_0}$ covered by unstable elements is not enough to make them sensitive to the boundaries of the propagation volume. This can be seen clearly in table A.4 where we find the rest frame lifetimes and corresponding values of L for some β radioactive nuclei at two different energies as presented in Donato et al. (2002). Of course the quoted distances strongly depend on the diffusion coefficient, which itself depends on the transport model used, but

Z	Nucleus	Daughter	$t_{1/2}^{\mathrm{unit}}$(error)
4	${}^{7}_{4}$Be	${}^{7}_{3}$Li	53.29^{d}(0.07)
18	${}^{37}_{18}$Ar	${}^{37}_{17}$Cl	35.04^{d}(0.04)
20	${}^{41}_{20}$Ca	${}^{41}_{19}$K	103^{kyr}(4)
22	${}^{44}_{22}$Ti	$({}^{44}_{21}\mathrm{Sc} \overset{\beta+}{\rightarrow}){}^{44}_{20}\mathrm{Ca}$	49^{yr}(3)
23	${}^{49}_{23}$V	${}^{49}_{22}$Ti	330^{d}(15)
24	${}^{48}_{24}$Cr[a]	$({}^{48}_{23}\mathrm{V} \overset{\beta+}{\rightarrow}){}^{48}_{22}\mathrm{Ti}$	21.56^{h}(0.03)[b]
24	${}^{51}_{24}$Cr	${}^{51}_{23}$V	27.702^{d}(0.004)
25	${}^{53}_{25}$Mn	${}^{53}_{24}$Cr	3.74^{Myr}(0.04)
26	${}^{55}_{26}$Fe	${}^{55}_{25}$Mn	2.73^{yr}(0.03)
27	${}^{57}_{27}$Co	${}^{57}_{26}$Fe	271.79^{d}(0.09)
28	${}^{59}_{28}$Ni[c]	${}^{59}_{27}$Co	80^{kyr}(11)

[a]This nucleus has an alowed β transition, but contrary to ^{54}Mn and ^{56}Ni, it has not been studied recently, so that we can set it as a pure K-capture decay.

[b]In this two step reaction, the second transition $^{48}\mathrm{V} \overset{\beta+}{\rightarrow} {}^{48}$Ti has a half lifetime grater than the first one (15.9735^{d}(0.0025). Nevertheless, this second reaction can be taken as immediate, because of its β nature. We thus can consider this second element as a ghost. Finally, only the first reaction ($^{48}\mathrm{Cr} \rightarrow ^{48}$ V) enters the decay rate.)

[c]This nucleus has a β decay, but with $\tau_{1/2} > 100$ Gyr, thus it is sufficient to take into account only the K-capture channel.

Table A.2: Pure K-capture isotopes from Donato et al. (2002)

even if the values were true for any model, the conclusion that the halo height cannot be constrained from radioactive instable isotopes is incorrect: Together with the transport parameters the halo height determines the CR flux and this way the escape time of CRs, which is measureable via the ratio of radioactive stable to instable isotopes.

Z	Nucleus	Daughter (EC)	$t_{1/2}^{\rm unit}$(error)	Daughter (β)	$t_{1/2}^{\rm unit}$(error)
13	$^{26}_{13}$Al	$^{26}_{12}$Mg	$4.08^{\rm Myr}(0.15)$	$^{26}_{12}$Mg	$0.91^{\rm Myr}(0.04)$
17	$^{36}_{17}$Cl	$^{36}_{16}$S	$15.84^{\rm Myr}(0.11)$	$^{36}_{18}$Ar	$0.307^{\rm Myr}(0.002)$
25	$^{54}_{25}$Mn	54_{24}Cr	$312.3^{\rm d}(0.4)$	$^{54}_{26}$Fe	$0.494^{\rm Myr}(0.006)$
28	$^{56}_{28}$Ni	$(^{56}_{27}{\rm Co} \stackrel{\beta+}{\rightarrow})^{56}_{26}{\rm Fe}^{a}$	$6.075^{\rm d}(0.02)$	$(^{56}_{27}{\rm Co} \stackrel{\beta+}{\rightarrow})^{56}_{26}{\rm Fe}$	$0.051^{\rm Myr}(0.022)$

[a] $^{56}_{26}Co$ decays via electronic capture (80%) and β^+ (20%). Since the half life time for electronic capture is of the order of two months, one cann assume that the only effective channel is β-decay. Note that these values are taken from Goldman (1982). More recent references Audi (1997) or nuclear charts on the web are ignored because they give either pure β channel or pure electronic capture channels.

Table A.3: Mixed K-captire and β-decay isotopes from Donato et al. (2002)

	τ_0(Myr)	1 GeV/nuc	10 GeV/nuc
^{10}Be	2.17	220 pc	950pc
^{26}Al	1.31	110 pc	470pc
^{36}Cl	0.443	56 pc	250 pc

Table A.4: Propagation distance for unstable nuclei. From Donato et al. (2002).

Appendix B

Crank-Nicholson coefficents

B.1 Crank-Nicholson coefficients for R-dependent convection

The propagation method used in GALPROP can be found in Strong & Moskalenko (1998). For the numerical solution of the transport equation the Crank-Nicholson implicit method is used Press et al. (1992). Following the notation of the GALPROP explanatory supplement Strong & Moskalenko (2006) we find the Crank-Nicholson coefficients for R-dependent convection in z-direction to be

$$\frac{\alpha_1}{\Delta t} = \frac{V(R_j, z_{i-1})}{z_i - z_{i-1}}, \frac{\alpha_2}{\Delta t} = \frac{V(R_j, z_i)}{z_i - z_{i-1}}, \frac{\alpha_3}{\Delta t} = 0 \tag{B.1}$$

for $z > 0$ and

$$\frac{\alpha_1}{\Delta t} = 0, \frac{\alpha_2}{\Delta t} = \frac{V(R_j, z_i)}{z_{i+1} - z_i}; \frac{\alpha_3}{\Delta t} = \frac{V(R_j, z_{i+1})}{z_{i+1} - z_i} \tag{B.2}$$

for $z < 0$.

For transport in momentum space the coefficients read

$$\frac{\alpha_1}{\Delta t} = 0$$

$$\frac{\alpha_2}{\Delta t} = \frac{V(R_j, z_{i+1}) - V(R_j, z_i)}{z_{i+1} - z_i} \frac{p_i}{3(p_{i+1} - p_i)} \tag{B.3}$$

$$\frac{\alpha_3}{\Delta t} = \frac{V(R_j, z_{i+1}) - V(R_j, z_i)}{z_{i+1} - z_i} \frac{p_{i+1}}{3(p_{i+1} - p_i)}$$

for $z > 0$ and

$$\frac{\alpha_1}{\Delta t} = 0$$

$$\frac{\alpha_2}{\Delta t} = \frac{V(R_j, z_{i-1}) - V(R_j, z_i)}{z_i - z_{i-1}} \frac{p_i}{3(p_{i+1} - p_i)} \tag{B.4}$$

$$\frac{\alpha_3}{\Delta t} = \frac{V(R_j, z_{i-1}) - V(R_j, z_i)}{z_i - z_{i-1}} \frac{p_{i+1}}{3(p_{i+1} - p_i)}$$

for $z < 0$.

B.2 Crank-Nicholson coefficients for anisotropic diffusion

Following the notation of the GALPROP explanatory supplement Strong & Moskalenko (2006) we find the Crank-Nicholson coefficients for the R and z dependent diffusion coefficients $D_{RR}(R, z, p)$ and $D_{zz}(R, z, p)$ to be

$$\frac{\alpha_1}{\Delta t} = \frac{D_{RR}(R_i, z_j, p_j)}{(R_{i+1} - R_i)(R_i - R_{i-1})} - \frac{D_{RR}(R_i, z_j, p_j)}{R_i(R_i - R_{i-1})} - \\ - \frac{D_{RR}(R_{i+1}, z_j, p_j) - D_{RR}(R_{i-1}, z_j, p_j)}{(R_{i+1} - R_{i-1})^2},$$

$$\frac{\alpha_2}{\Delta t} = \frac{D_{RR}(R_i, z_j, p_j)}{(R_{i+1} - R_i)(R_{i+1} - R_i)} + \frac{D_{RR}(R_i, z_j, p_j)}{(R_{i+1} - R_i)(R_i - R_{i-1})}, \tag{B.5}$$

$$\frac{\alpha_3}{\Delta t} = \frac{D_{RR}(R_{i+1}, z_j, p_j) - D_{RR}(R_{i-1}, z_j, p_j)}{(R_{i+1} - R_{i-1})^2} - \\ - \frac{D_{RR}(R_i, z_j, p_j)}{R_i(R_{i+1} - R_{i-1})} + \frac{D_{RR}(R_{i+1}, z_j, p_j)}{(R_{i+1} - R_i)^2}$$

for transport in R direction and

$$\frac{\alpha_1}{\Delta t} = \frac{D_{zz}(R_j, z_i, p_j)}{(z_{i+1} - z_i)(z_i - z_{i-1})} - \frac{D_{zz}(R_j, z_{i+1}, p_j) - D_{zz}(R_j, z_{i-1})}{(z_{i+1} - z_{i-1})^2},$$

$$\frac{\alpha_2}{\Delta t} = \frac{D_{zz}(R_j, z_i, p_j)}{(z_{i+1} - z_i)^2} - \frac{D_{zz}(R_j, z_i, p_j)}{(z_{i+1} - z_i)(z_i - z_{i-1})}, \tag{B.6}$$

$$\frac{\alpha_3}{\Delta t} = \frac{D_{zz}(R_j, z_{i+1}, p_j) - D_{zz}(R_j, z_{i-1}, p_j)}{(z_{i+1} - z_{i-1})^2} + \frac{D_{zz}(R_j, z_i, p_j)}{(z_{i+1} - z_i)^2}$$

for transport in z direction. In the limit of an equidistant grid and constant diffusion our Crank-Nicholson coefficients agree with those used in Strong & Moskalenko (2006) except for a factor 2 in α_1 and α_3 for transport along R. However, deriving these coefficients for the case of an equidistant grid and constant diffusion from their Eq. 25 we find

$$\frac{\alpha_1}{\Delta t} = D_{RR} \frac{R_i - \Delta R}{R_i (\Delta R)^2},$$

and

$$\frac{\alpha_3}{\Delta t} = D_{RR} \frac{R_i + \Delta R}{R_i (\Delta R)^2}$$

in agreement with our coefficients. The coefficients for diffusive reacceleration remain unchanged, because here only derivatives in momentum space occur. However, if $D_{zz} > D_{RR}$ one has to keep in mind that v_α has to be considered an effective parameter, scaled with the anisotropy in diffusion, e.g. $\bar{v}_\alpha = v_\alpha \cdot \sqrt{D_{zz}/D_{RR}}$

Appendix C

Halo Parameters

The halo profile used for the analysis of the EGRET and Fermi-LAT data is taken from Sander (2005). It can be parameterized by a spherical halo with the superposition of two Gaussian rings at radius R_n with a width $\sigma_{R,n}$. The ring density above the plane is assumed to decrease exponetially with a decay constant $\sigma_{z,n}$. The total halo profile can be written as

$$\rho_\chi(\vec{r}) = \rho_0 \cdot \left(\frac{\tilde{r}}{r_0}\right)^{-\gamma} \left[\frac{1+\left(\frac{\tilde{r}}{a}\right)^\alpha}{1+\left(\frac{r_0}{a}\right)^\alpha}\right]^{\frac{\gamma-\beta}{\alpha}} + \sum_{n=1}^{2} \rho_n \exp\left(-\frac{(\tilde{r}_{gc,n} - R_n)^2}{2 \cdot \sigma_{R,n}^2} - \left|\frac{z}{\sigma_{z,n}}\right|\right) \tag{C.1}$$

with

$$\tilde{r} = \sqrt{x^2 + \frac{y^2}{\varepsilon_{xy}^2} + \frac{z^2}{\varepsilon_z^2}}, \qquad \tilde{r}_{gc,n} = \sqrt{x_{(n)}^2 + \frac{y_{(n)}^2}{\varepsilon_{xy,n}^2}}, \tag{C.2}$$

and the excentricities ε_{xy} and ε_z of the triaxial halo profile and the excentricities $\varepsilon_{xy,n}$ of the elliptical rings. Additional degrees of freedom are the angles with respect to the axis earth - galactic center of the halo ϕ_{gc} and of the rings ϕ_n, i.e. each component has its own coordinate system which is rotated around the z-axis.

The radial width of the outer ring is taken to be different for the inner and outer side as expected from the infall of a dwarf galaxy. Since infalling matter has an angular momentum with respect to the GC it cannot reach the GC. A Gaussian profile still has a non vanishing density in the galactic center so the shape was modified to fall off to zero within a distance

d_n. The fall off is parametrized by two quadratic functions

$$\rho(r) = \begin{cases} \frac{4\rho_n}{d_n^2} \cdot (r - (R_n - d_n))^2 & \text{for } (R_n - d_n) < r < (R_n - d_n/2), \\ \rho_n - \frac{4\rho_n}{d_n^2} \cdot (r - R_n)^2 & \text{for } (R_n - d_n/2) < r < R_n \end{cases} \tag{C.3}$$

The parameters of the halo profile used in this work can be found in Table C.1 and are those of Sander (2005). Details on the implementation of the halo profile and the fit can be found in Sander (2005) and de Boer et al. (2005).

Parameter	PISO with rings
ρ_0 [GeV cm^{-3}]	0.5[a] (0.725[b])
ρ_{tot} at earth [GeV cm^{-3}]	1.2
r_0 [kpc]	8.3
α	2
β	2
γ	0
a [kpc]	5
ε_{xy}	0.8
ε_z	0.75
ϕ_{gc} [°]	90
ρ_1 [GeV cm^{-3}]	4.5
R_1 [kpc]	4.15
$\sigma_{r,1}$ [kpc]	4.15
$\sigma_{z,1}$ [kpc]	0.17
$\varepsilon_{xy,1}$	0.8
ϕ_1 [°]	-70
M_1 [M$_\odot$]	$9.3 \cdot 10^9$
ρ_2 [GeV cm^{-3}]	1.85
R_2 [kpc]	12.9
$\sigma_{r,2}$ [kpc]	3.3
d_2 [kpc]	4
$\sigma_{z,2}$ [kpc]	1.7
$\varepsilon_{xy,2}$	0.95
ϕ_2 [°]	-20
M_2 [M$_\odot$]	$8.5 \cdot 10^{10}$
boost-factor (GALPROP)	10[c]
boost-factor (halofitter, rings)	8.61[d]
boost-factor (halofitter, no rings)	10.69[e]

[a]For the halo profile from Sander (2005) used in GALPROP

[b]For the halo profile used in the fits performed with the halofitter presented in this work. The difference in the local halo density results from the fact that the local halo denisty is normalized to the rotation curve at the position of the Sun. Different from Sander (2005) we here use the slightly smaller velocity of 244 km/s (Gillessen et al., 2009; Reid & Brunthaler, 2004).

[c]This boost factor refers to ρ_0=0.5 [GeV cm^{-3}]

[d]This boost factor refers to ρ_0=0.725 [GeV cm^{-3}]

[e]This boost factor refers to ρ_0=0.725 [GeV cm^{-3}]

Table C.1: Fit results for the pseudo-isothermal profile with rings; for the calculation of the local density the position of the earth was assumed to be 0.1 kpc above the galactic plane. From Sander (2005).

Appendix D

Magnetic Mirrors and Trapped CRs

In section 3.5.4 we introduced the idea of CR trapping by molecular cloud complexes. If magnetic reflection on molecular clouds is indeed efficient, the CR interaction rate with the H_2 component of the ISM would be greatly reduced. Trapping by MCCs cannot entirely be modelled by diffusion. At any point in time CRs consist of a passing and a trapped fraction. The passing fraction is taken care of by the diffusion tensor, the trapped fraction is not subject to a purely diffusive movement. Although trapped particles scatter on magnetic turbulences the scattering always occurs under an angle of π, thus allowing for momentum losses, fragmentation and radioactive decays in the atomic phase of the gas without any change of the spatial CR distribution. It is entirely possible that the trapping efficiency of MCCs is very small. Even in this case still an effective reflection from the high field regions in MCs is expected, because the spectral analysis of the positron annihilation line strongly supports reflection of CRs by MCs (see section 3.5.4). A detailed description of MCC trapping can be arbitrary complicated: The time CRs are trapped inside a MCC and the depth CRs penetrate into MCs are basically unknown. In addition MCCs are moving in the ISM and can thus contribute signififanctly to CR reacceleration (Zirakashvili, 1999) (see section 2.3.2). Without any assumptions about the efficiency of trapping by MCCs, we can estimate the maximum impact of trapping by molecular clouds by assuming that the H_2 distribution does not contribute the secondary production, which means that CRs are totally excluded from MCs. Figure D.1 shows the contribution of the H_2 component to B/C and $^{10}Be/^{9}Be$.

An intresting feature of a model with reflection by MCs is the fact that the soft γ-ray gradient problem dissappears: From Fig. 3.2 it is clear that the peak in the gas distribution and consequently the γ-ray emissivity is due to molecular hydrogen. If CRs are excluded

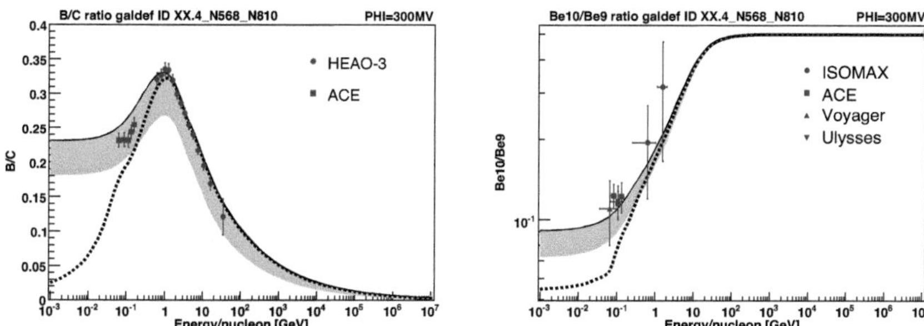

Figure D.1: The reduction in B/C (**left**) and $^{10}Be/^{9}Be$ (**right**) exptected in an aPM with complete exclusion of CRs from MCs.

from this phase of the ISM by magnetic reflection, the CRs interaction rate with H_2 is zero and consequently the gradient in diffuse γ-rays will be flat enough to be compatible with the SNR distribution without the help of convection.

Modelling Magnetic Trapping Assuming a homogeneous gas distribution the interaction and decay rate is just proportional to the average length of the path CRs traveled: $1/\tau_{i,d} \sim s$. In the presence of magnetic traps this pathlength is elongated by the additional pathlength CRs lay back in the trapping regions $1/\tau'_{i,d} \sim s + s_t$. In a diffusion model with the interaction rate defined by the diffusion coefficient (and other transport parameters), the interaction and decay rate $1/\tau_{i,d}$ has to be replaced by the increased rate $1/\tau'_{i,d} = G_{MCC} \cdot r_{i,d}$ with $G_{MCC} = 1 + s/s_t$. This way, in the presence of trapping the fragmentation and decay rate, momentum losses and secondary production rate in the molecular component of the gas remain unchanged and while the corrresponding processes in the atomic component are enhanced by a factor G_{MCC}.

The trapping efficiency G_{MCC} has another physical meaning: it is given by the ratio of the maximum field strength B_{max} a CR encounters at the mirror point and the typical field strength in the ISM B_{ISM}. Since B_{ICM}/B_{max} is only weakly constrained G_{MCC} is essentially a free parameter. Chandran (2000) considered values of up to 200 for G_{MCC}. Note, that the above is still a very simple estimate. The trapping efficiency G_{MCC} might feature and energy dependence due to the energy dependent diffusive scattering length and in addition a reacceleration process is expected to occur if the traps are moving.

Bibliography

Abdo, A. A., Ackermann, M., Ajello, M., et al. 2009, Phys. Rev. Lett., 102, 181101

Abdo, A. A., Ackermann, M., Ajello, M., et al. 2009a, Phys. Rev. Lett., 102, 181101

Abdo, A. A., Ackermann, M., Ajello, M., et al. 2009b, Astrophys. J., 703, 1249

Abraham, J., Aglietta, M., Aguirre, I. C., et al. 2004, Nuclear Instruments and Methods in Physics Research A, 523, 50

Adriani, O., Barbarino, G. C., Bazilevskaya, G. A., et al. 2009a, Nature, 458, 607

Adriani, O., Barbarino, G. C., Bazilevskaya, G. A., et al. 2009b, Phys. Rev. Lett., 102, 051101

Aharonian, F., Akhperjanian, A. G., Bazer-Bachi, A. R., et al. 2007, Astron. Astrophys., 464, 235

Aharonian, F. A., Atoyan, A. M., & Voelk, H. J. 1995, Astron. Astrophys., 294, L41

Akhiezer, A. I. 1975, Plasma electrodynamics - Vol.1: Linear theory; Vol.2: Non-linear theory and fluctuations, ed. Akhiezer, A. I. (International Series of Monographs in Natural Philosophy, Oxford: Pergamon Press, 1975)

Alcaraz, J., Alpat, B., Ambrosi, G., et al. 2000, Phys. Lett., B490, 27

Amaldi, U., de Boer, W., & Furstenau, H. 1991, Phys. Lett., B260, 447

Anderson, C. D. 1933, Phys. Rev., 43, 491

Arkani-Hamed, N., Finkbeiner, D. P., Slatyer, T. R., & Weiner, N. 2009, Phys. Rev. D, 79, 015014

Asakimori, K., Burnett, T. H., Cherry, M. L., et al. 1998, Astrophys. J., 502, 278

Ascasibar, Y., Jean, P., Boehm, C., & Knoedlseder, J. 2006, Mon. Not. Roy. Astron. Soc., 368, 1695

Asztalos, S. J., Rosenberg, L. J., van Bibber, K., Sikivie, P., & Zioutas, K. 2006, Ann. Rev. Nuc. Part. Sc., 56, 293

Audi, G. 1997, Nucl. Phys. A, 624, 1

Babcock, H. W. 1939, Lick Observatory bulletin, 41, 498

Basini, G. 1999, in International Cosmic Ray Conference, Vol. 3, International Cosmic Ray Conference, 77–+

Bell, A. R. 1978, Mon. Not. Roy. Astron. Soc., 182, 147

Berezinsky, V., Dokuchaev, V., & Eroshenko, Y. 2008, Phys. Rev. D, 77, 083519

Bergström, L. 2009, New Journal of Physics, 11, 105006

Bergstrom, L., Edsjo, J., Gustafsson, M., & Salati, P. 2006, JCAP, 0605, 006

Bergström, L., Edsjö, J., & Zaharijas, G. 2009, Phys. Rev. Lett., 103, 031103

Bertone, G., Hooper, D., & Silk, J. 2005, Phys. Rep., 405, 279

Bian, N. H. & Garcia, O. E. 2005, Physics of Plasmas, 12, 042307

Blandford, R. & Ostriker, J. 1978, Astrophys. J., 221, L29

Blasi, P. 2009, Phys. Rev. Lett., 103, 051104

Blasi, P. & Serpico, P. D. 2009, Phys. Rev. Lett., 103, 081103

Boehm, C., Hooper, D., Silk, J., Casse, M., & Paul, J. 2004, Phys. Rev. Lett., 92, 101301

Boezio, M., Carlson, P., Francke, T., et al. 2000, Astrophys. J, 532, 653

Bradt, H. L. & Peters, B. 1950, Phys. Rev., 80, 943

Breitschwerdt, D. 2008, Nature, 452, 826

Breitschwerdt, D. & de Avillez, M. A. 2006, in Invited Paper at IAU Symposium 237 "Triggered Star Formation in a Turbulent ISM", Bruce Elmegreen and Jan Palous (eds.)

Breitschwerdt, D., Dogiel, V. A., & Völk, H. J. 2002, Astron. Astrophys., 385, 216

Breitschwerdt, D., McKenzie, J. F., & Voelk, H. J. 1991, Astron. Astrophys., 245, 79

Bronfman, L. 1988, Astrophys. J, 324, 248

Burton, W. B. 1988, The structure of our Galaxy derived from observations of neutral hydrogen, ed. K. I. Kellermann & G. L. Verschuur (Galactic and extragalactic radio astronomy (2nd edition) (A89-40409 17-90). Berlin and New York, Springer-Verlag, 1988, p. 295-358.), 295–358

Case, G. & Bhattacharya, D. 1996, Astron. Astrophys. Suppl., 120, C437+

Cesarsky, C. J. 1980, Ann. Rev. Astron. Astrophys., 18, 289, and references therein.

Chamseddine, A. H., Arnowitt, R. L., & Nath, P. 1982, Phys. Rev. Lett., 49, 970

Chandran, B. D. G. 2000, Astrophys. J., 529, 513

Chandran, B. D. G. 2001, Space Sci. Rev., 99, 271

Chang, J., Adams, J. H., Ahn, H. S., et al. 2008, Nature, 456, 362

Chowdhury, D., Jog, C. J., & Vempati, S. K. 2009, ArXiv e-prints:0909.1182

Codino, A. & Plouin, F. 2007, ArXiv e-print:0701521

Connell, J. & et al. 1997, in International Cosmic Ray Conference, Vol. 3, International Cosmic Ray Conference, 397–+

Connell, J. J. 1998, Astrophys. J. Lett., 501, L59+

Connell, J. J., DuVernois, M. A., & Simpson, J. A. 1998, ApJL, 509, L97

Cordes, M. e. a. 1991, Nature, 354, 121

Coutu, S., Barwick, S. W., Beatty, J. J., et al. 1999, Astropart. Phys., 11, 429

Cox, P., E., K., & Mezger, P. G. 1986, Astron. Astrophys, 155, 380

Davis, A. e. a. 2000, in AIP Conf. Proc. 528, Acceleration and Transport of Energetic Particles Observed in the Heliosphere (ACE-2000), eds. R. A. Mewaldt et al. (New York: AIP), 421 de Boer, W. 1994, Prog. Part. Nucl. Phys., 33, 201

de Boer, W. 2009, ArXiv e-prints:0910.2601, submitted to the SUSY09 proceedings

de Boer, W., Sander, C., Zhukov, V., Gladyshev, A. V., & Kazakov, D. I. 2005, Astron. Astrophys., 444, 51

de Boer, W., Sander, C., Zhukov, V., Gladyshev, A. V., & Kazakov, D. I. 2006, Physics Letters B, 636, 13

De Marco, D., Blasi, P., & Stanev, T. 2007, JCAP, 0706, 027

de Nolfo, G. A., Moskalenko, I. V., Binns, W. R., et al. 2006, Advances in Space Research, 38, 1558

Dermer, C. D. 1985, Astrophys. J, 295, 28

Diehl, R., Halloin, H., Kretschmer, K., et al. 2006, Nature, 439, 45

Diemand, J., Kuhlen, M., & Madau, P. 2007, Astrophys. J., 262, 657

Diffuse and Molecular Clouds Science Working Group Fermi-LAT. 2009, http://fermi.gsfc.nasa.gov/ssc/data/access/lat/ring_for_FSSC_final4.pdf

Dobler, G. & Finkbeiner, D. P. 2008, Astrophys. J., 680, 1222

Dobler, G., Finkbeiner, D. P., Cholis, I., Slatyer, T. R., & Weiner, N. 2009, ArXiv e-prints:0910.4583, submitted to Astrophys. J.

Dogiel, V. A., Gurevich, A. V., & Zybin, K. P. 1994, Astron. Astrophys., 281, 937

Dogiel, V. A. & Gurevich, A. V. e. 1993, in International Cosmic Ray Conference, Vol. 2, International Cosmic Ray Conference, 275–+

Donato, F., Maurin, D., & Tailet, R. 2002, Asron. Astrophys, 381, 539

Donato, F., Maurin, D., & Taillet, R. 2002, Astron. Astrophys., 381, 539

Downes, D. & Guesten, R. 1982, Mitteilungen der Astronomischen Gesellschaft Hamburg, 57, 207

Drury, L. O. 1983, Reports on Progress in Physics, 46, 973

DuVernois, M. A. 1997, Astrophys. J., 481, 241

DuVernois, M. A., Barwick, S. W., Beatty, J. J., et al. 2001, Astrophys. J., 559, 296

DuVernois, M. A., Simpson, J. A., & Thayer, M. R. 1996, Astron. Astrophys., 316, 555

Edsjo, J., Schelke, M., Ullio, P., & Gondolo, P. 2003, JCAP, 0304, 001

Engelmann, J. e. a. 1985, Astron. Astrophys., 148, 12

Engelmann, J. J., Ferrando, P., Soutoul, A., Goret, P., & Juliusson, E. 1990, Astron. Astrophys., 233, 96

Everett, J. E., Zweibel, E. G., Benjamin, R. A., et al. 2008, Astrophys. J., 674, 258

Fermi, E. 1949, Phys. Rev., 75, 1169

Ferrara, S., Wess, J., & Zumino, B. 1974, Phys. Lett., B51, 239

Freier, P. S., Young, J. S., & Waddington, C. J. 1980, Astrophys. J, 240, L53

Frisch, P. C. 2009, Space Science Reviews, 143, 191

Fuchs, B., Breitschwerdt, D., de Avillez, M. A., Dettbarn, C., & Flynn, C. 2006, Mon. Not. Roy. Astron. Soc., 373, 993

Gaisser, T. K. 1990, Cosmic rays and particle physics, ed. Gaisser, T. K. (Cambridge and New York, Cambridge University Press, 1990, 292 p.)

Gaisser, T. K. 2001, in American Institute of Physics Conference Series, Vol. 558, American Institute of Physics Conference Series, ed. F. A. Aharonian & H. J. Völk, 27–42

Gaisser, T. K. 2007, in Energy Budget in the High Energy Universe, ed. K. Sato & J. Hisano (World Scientific Publishing Co., Pte. Ltd., Singapore), 45–+

Garcia-Munoz, M., Mason, G. M., & Simpson, J. A. 1975, Astrophys. J. Lett., 201, L145

Garcia-Munoz, M., Mason, G. M., & Simpson, J. A. 1977, Astrophys. J, 217, 859

Gast, H. & Schael, S. 2009, in Proceedings of the 31st ICRC, Lodz, ICRC

George, J. S., Lave, K. A., Wiedenbeck, M. E., et al. 2009, Astrophys. J., 698, 1666

Gillessen, S., Eisenhauer, F., Trippe, S., et al. 2009, Astrophys. J., 692, 1075

Ginzburg, V. 1979, Theoretical Physics and Astrophysics (Pergamon Press, Oxford)

Ginzburg, V. L., Dogiel, V. A., Berezinsky, V. S., Bulanov, S. V., & Ptuskin, V. S. 1990, Astrophysics of cosmic rays (Amsterdam, Netherlands: North-Holland (1990) 534 p)

Ginzburg, V. L., Khazan, I. M., & Ptuskin, V. S. 1980, Astrophys. Sp. Sc., 68, 295

Gleeson, L. J. & Axford, W. I. 1968, Astrophys. J., 154, 1011

Goldman, D. T. 1982, American Institute of Physics Handbook, third edition, par. 8

Gondolo, P., Edsjö, J., Ullio, P., et al. 2004, JCAP, 7, 8

Gordon, M. A. & Burton, W. B. 1976, Astrophys. J, 208, 346

Grasso, D., Profumo, S., Strong, A. W., et al. 2009, Astropart. Phys., 32, 140

Grevesse, N., Noels, A., & Sauval, A. J. 1996, ASP Conf. Series, 99, 117

Guessoum, N., Jean, P., & Gillard, W. 2005, Astron. Astrophys., 436, 171

H. E. S. S. Collaboration: F. Aharonian. 2009, ArXiv e-prints:0905.0105

Hagen, F. A., Fischer, A. J., & Ormes, J. F. 1977, Astrophys. J, 212, 262

Hams, T., Barbier, L. M., Bremerich, M., et al. 2004, Astrophys. J., 611, 892

Han, J.-L. 2004, astro-ph/0402170

Haslam, C. G. T., Salter, C. J., Stoffel, H., & Wilson, W. E. 1982, Astron. Astrophys. Suppl., 47, 1

Haug, E. 1975, Z. Naturforsch., 30a, 1546

Heiles, C. 1996, Astrophys. J., 462, 316

Hess, V. F. 1912, Phys. Zeitschrift, 13, 1084

Hooper, D., Blasi, P., & Dario Serpico, P. 2009, JCAP, 1, 25

Hooper, D., Finkbeiner, D. P., & Dobler, G. 2007, Phys. Rev., D76, 083012

Hooper, D. & Wang, L.-T. 2004, Phys. Rev., D70, 063506

Hou, L. G., Han, J. L., & Shi, W. B. 2009, Astron. Astrophys., 499, 473

Hunter, S. D., Bertsch, D. L., Catelli, J. R., et al. 1997, Astrophys. J, 481, 205

Jean, P., Gillard, W., Marcowith, A., & Ferrière, K. 2009, ArXiv e-prints:0909.4022

Jean, P., Knödlseder, J., Gillard, W., et al. 2006, Astron. Astrophys., 445, 579

Jokipii, J. R. 1976, Astrophys. J., 208, 900

Jones, F. C. 1965, Phys. Rev, 77, 54

Jungman, G., Kamionkowski, M., & Griest, K. 1996, Phys. Rep., 195, 267

Kalberla, P. M. W., Dedes, L., Kerp, J., & Haud, U. 2007, Astron. Astrophys., 469, 511

Kennel, C. & Engelmann, F. 1966, Phys. Fluids, 9, 2377

Knödlseder, J., Jean, P., Lonjou, V., et al. 2005, Astron. Astrophys., 441, 513

Koch, H. & Motz, J. W. 1959, Rev. Mod. Phys, 31, 920

Kolb, E. W. & Slansky, R. 1984, Phys. Lett. B, 135, 378

Komatsu, E., Dunkley, J., Nolta, M. R., et al. 2009, Astrophys. J. Suppl., 180, 330

Lagage, P. O. & Cesarsky, C. J. 1983a, Astron. Astrophys., 118, 223

Lagage, P. O. & Cesarsky, C. J. 1983b, Astron. Astrophys., 125, 249

Launhardt, R., Zylka, R., & Mezger, P. G. 2002, Astron. Astrophys., 384, 112

Leske, R. A. 1993, Astrophys. J, 405, 567

Levenson, N. A., Graham, J. R., Aschenbach, B., et al. 1997, Astrophys. J., 484, 304

Lodders, K. 2003, Astrophys. J., 591, 1220

Longair, M. 1992, High Energy Astrophysics (Cambridge University Press), reprinted with corrections 2004

Lorimer, D. R. 2004, in IAU Symposium, Vol. 218, Young Neutron Stars and Their Environments, ed. F. Camilo & B. M. Gaensler, 105–+

Lukasiak, A. 1999, in International Cosmic Ray Conference, Vol. 3, International Cosmic Ray Conference, 41–+

Lukasiak, A. e. a. 1994, Astrophys. J, 423, 426

Mannheim, K. & Schlickeiser, R. 1994, Astron. Astrophys, 286, 938

Martin, N. F., Ibata, R. A., Bellazzini, M., et al. 2004, Mon. Not. Roy. Astron. Soc., 348, 12

Martínez-Delgado, D., Pohlen, M., Gabany, R. J., et al. 2009, Astrophys. J., 692, 955

Mathewson, D. S. & Ford, V. L. 1970, MNRAS, 74, 139

Maurin, D., Donato, F., Taillet, R., & Salati, P. 2001, Astrophys. J., 555, 585

Meyer, J.-P., Drury, L. O., & Ellison, D. C. 1997, Astrophys. J, 487, 182

Mizuno, T. 2001, Nucl. Sci. Symp. Conf. Rec. IEEE, 1, 842

Moskalenko, I. & Jourdain, E. 1997, Astron. Astrophys, 325, 401

Moskalenko, I. V., Porter, T. A., & Strong, A. W. 2006, Astrophys. J. Lett., 640, L155

Moskalenko, I. V., Strong, A. W., & Reimer, O. 1998, Astron. Astrophys., 338, L75

Neddermeyer, S. H. & Anderson, C. D. 1937, Phys. Rev., 51, 884

Nilsen, B. S. 1998, in Eighteenth Texas Symposium on Relativistic Astrophysics, ed. A. V. Olinto, J. A. Frieman, & D. N. Schramm, 438–+

Nomura, Y. & Thaler, J. 2009, Phys. Rev. D, 79, 075008

Orito, S., Maeno, T., Matsunaga, H., et al. 2000, Phys. Rev. Lett., 84, 1078

Panov, A. D., Adams, J. H., Ahn, H. S., et al. 2006, in Proc. 29th All-Russian Cosmic Ray Conference, Moscow, Russia

Peñarrubia, J., Martínez-Delgado, D., Rix, H. W., et al. 2005, Astrophys. J., 626, 128

Peccei, R. D. & Quinn, H. R. 1977, Phys. Rev. Lett., 38, 1440

Perkins, D. H. 1947, Nature, 159, 126

Phno, H. & Shibata, S. 1993, MNRAS, 262, 953

Phyllips, S. e. a. 1981, Astron. Astrophys., 103, 405

Pierre Auger Collaboration. 2007, Science, 318, 938

Porter, T. A. 2009, in International Cosmic Ray Conference, 2009

Porter, T. A. e. a. 2005, in International Cosmic Ray Conference, Vol. 4, International Cosmic Ray Conference, 77–+

Prantzos, N. 2006, Astron. Astrophys., 449, 869

Press, W. H., Teukolsky, S. A., Vetterling, W. T., & Flannery, B. P. 1992, Numerical recipes in FORTRAN. The art of scientific computing (Cambridge: University Press, —c1992, 2nd ed.)

Profumo, S. 2008, arXiv:astro-ph/0812.4457, submitted to Phys.Rev.D

Ptuskin, V. S. & Khazan, Y. M. 1976, Astrophysics, 12, 81

Ptuskin, V. S., Moskalenko, I. V., Jones, F. C., Strong, A. W., & Zirakashvili, V. N. 2006, Astrophys. J., 642, 902

Rand, R. J. & Lyne, A. G. 1994, MNRAS, 268, 497

Reid, M. J. & Brunthaler, A. 2004, Astrophys. J., 616, 872

Rochester, G. D. & Butler, C. C. 1947, Nature, 160, 855

Sander, C. 2005, Interpretation des Überschusses in diffuser galaktischer Gamma-Strahlung oberhalb 1GeV als Annihilationssignal dunkler Materie (Univ. Karlsruhe, PhD thesis, IEKP-KA/2005-12)

Sattin, F. 2008, Phys. Lett. A, 372, 3941

Schlickeiser, R. 2002, Cosmic ray astrophysics (Berlin, Germany, Springer)

Seljak, U., Slosar, A., & McDonald, P. 2006, JCAP, 10, 14

Serpico, P. D. 2009, Phys. Rev. D, 79, 021302

Servant, G. & Tait, T. M. P. 2003, Nucl. Phys. B, 650, 391

Shaviv, N. J., Nakar, E., & Piran, T. 2009, Phys. Rev. Lett., 103, 111302

Snowden, S. L., Egger, R., Freyberg, M. J., et al. 1997, Astrophys. J., 485, 125

Springel, V., White, S. D. M., Frenk, C. S., et al. 2008, Nature, 456, 73

Sreekumar, P., Bertsch, D. L., Dingus, B. L., et al. 1998, Astrophys. J., 494, 523

Stanev, T. 2004, High energy cosmic rays (Berlin, Germany, Springer)

Street, J. C. & Stevenson, E. C. 1937, Phys. Rev., 52, 1003

Strong, A. & Mattox, J. 1996, Astron. Astrophys., 308, L21

Strong, A. & Moskalenko, I. 2006, http://galprop.stanford.edu/manuals/manual.pdf

Strong, A. W. & Moskalenko, I. V. 1998, Astrophys. J., 509, 212

Strong, A. W. & Moskalenko, I. V. 2001, Adv. Space Res., 27, 4 717

Strong, A. W. & Moskalenko, I. V. 2009, in International Cosmic Ray Conference, Paper ID 0626

Strong, A. W., Moskalenko, I. V., & Ptuskin, V. S. 2007, Ann. Rev. Nucl. Part. Sci., 57, 285

Strong, A. W., Moskalenko, I. V., & Reimer, O. 2000, Astrophys. J., 537, 763

Strong, A. W., Moskalenko, I. V., & Reimer, O. 2000, Astrophys. J., 537, 763

Strong, A. W., Moskalenko, I. V., & Reimer, O. 2004a, Astrophys. J., 613, 962

Strong, A. W., Moskalenko, I. V., & Reimer, O. 2005, in American Institute of Physics Conference Series, Vol. 745, High Energy Gamma-Ray Astronomy, ed. F. A. Aharonian, H. J. Völk, & D. Horns, 585–590

Strong, A. W., Moskalenko, I. V., Reimer, O., Digel, S., & Diehl, R. 2004b, Astron. Astrophys., 422, L47

Tata, X. 1997, lectures given at 9th Jorge Andre Swieca Summer School: Particles and Fields, Sao Paulo, Brazil, 16-28 Feb 1997. ArXiv e-print:9706307

Taylor, J. H. & Cordes, J. M. 1993, Astrophys. J., 411, 674

Torii, S., Yamagami, T., Tamura, T., et al. 2008, arXiv:astro-ph/0809.0760, submitted to Astropart. Phys

Troland, T. H. & Heiles, C. 1986, Astrophys. J., 301, 339

Vallée, J. P. 2005, Astronom. J., 130, 569

van Milligen, B. P., Bons, P. D., Carreras, B. A., & Sanchez, R. 2005, European Journal of Physics, 26, 913

von Stickforth, J. 1961, Z. Physik, 164, 1

Wang, J. Z., Seo, E. S., Anraku, K., et al. 2002, Astrophys. J., 564, 244

Webber, W. R. 1987, Astron. Astrophys., 179, 277

Webber, W. R., Lezniak, J. A., Kish, J. C., & Simpson, G. A. 1977, Astrophys. Lett., 18, 125

Webber, W. R., Lukasiak, A., McDonald, F. B., & Ferrando, P. 1996, Astrophys. J., 457, 435

Weidenspointner, G., Knödlseder, J., Jean, P., et al. 2007, in ESA Special Publication, Vol. 622, ESA Special Publication, 25–+

Weidenspointner, G., Skinner, G., Jean, P., et al. 2008, Nature, 451, 159

Weinberg, S. 1978, Phys. Rev. Lett., 40, 223

Wess, J. & Zumino, B. 1974a, Phys. Lett., B49, 52

Wess, J. & Zumino, B. 1974b, Nucl. Phys., B70, 39

Wiedenbeck, M. E. 1985, ICRC 19, 2, 84

Wilczek, F. 1978, Phys. Rev. Lett., 40, 279

WiZard/CAPRICE Collaboration. 2001, Astrophys. J., 561, 787

Yanasak, N. E., Wiedenbeck, M. E., Mewaldt, R. A., et al. 2001, Astrophys. J., 563, 768

Yüksel, H., Kistler, M. D., & Stanev, T. 2009, Phys. Rev. Lett., 103, 051101

Zirakashvili, V. N. 1999, in Proc. 26th ICRC 4,439, Salt Lake City, USA

Zweibel, E. G. & Heiles, C. 1997, Nature, 385, 131

Zwicky, F. 1933, Helv. Phys. Acta, 110, 6

Printed by Books on Demand GmbH, Norderstedt / Germany